DESIGNING RENEWABLE ENERGY SYSTEMS

Léda Gerber

DESIGNING RENEWABLE ENERGY SYSTEMS

A LIFE CYCLE ASSESSMENT APPROACH

EPFL Press
Distributed by CRC Press

CRC Press
Taylor & Francis Group

Taylor and Francis Group, LLC
6000 Broken Sound Parkway, NW, Suite 300,
Boca Raton, FL 33487

Distribution and Customer Service
orders@crcpress.com

www.crcpress.com

Library of Congress Cataloging-in-Publication Data
A catalog record for this book is available from the Library of Congress.

This book is published under the editorial direction of Professor Christof Holliger (EPFL).

For the work described in this book, Léda Gerber was awarded the "EPFL Press Distinction",
an official prize discerned annually at the Ecole polytechnique fédérale de Lausanne (EPFL) and
sponsored by the Presses polytechniques et universitaires romandes. The Distinction is given to the
author of a doctoral thesis deemed to have outstanding editorial, instructive and scientific qualities;
the award consists of the publication of a book issued from their thesis.

Cover page: © cybercrisi, Fotolia.com

EPFL Press

is an imprint owned by the Presses polytechniques et universitaires romandes,
a Swiss academic publishing company whose main purpose is to publish the teaching
and research works of the Ecole polytechnique fédérale de Lausanne (EPFL).

Presses polytechniques et univeristaires romandes
EPFL – Rolex Learning Center
Post office box 119
CH-1015 Lausanne, Switzerland
E-mail: ppur@epfl.ch

www.epflpress.org

© 2014, First edition, EPFL Press
ISBN 978-2-940222-81-0 (EPFL Press)
ISBN 978-1-4987-1127-2 (CRC Press)
Printed in Italy

Foreword

This book is at the convergence of two worlds: life cycle environmental impact assessment and process system engineering. The first world starts from predefined scenarios to identify the best option, the second aiming at generating the best scenario from decision variables without a priori. It was therefore not an easy task to make these two scientific communities speak with one another, accepting and integrating each other's vocabulary, integrating concepts and proposing a unified methodology. Léda's work is therefore remarkable. She is demonstrating how life cycle impact assessment metrics and methods can be used in the process system engineering decision support. She is demonstrating how process system design methods can be used to generate pertinent scenarios for the life cycle environmental impact assessment. Extending the system boundaries she demonstrates how this method can be use to generate industrial ecology concepts.

In her thesis, Léda has developed and adapted the life cycle assessment methodology in order to represent the impact of the engineering decisions. This approach, known as parametric modeling in the life cycle assessment community, uses the conventional tools of flowsheeting and the design methods to calculate the impact of decisions with life cycle impact assessment metrics. By considering together impact assessment and sizing models, Léda is demonstrating how process design decisions will influence the environmental impacts. The method is particularly useful when developing innovative renewable energy conversion processes like biofuels, biorefineries, solar or geothermal plants. In contrast to LCA techniques that rely on the definition of a priori scenarios, the use of optimization techniques allows engineers to generate optimal scenarios based on the sustainability criteria like costs and environmental impact.

In this book, the detailed methodology is presented. It aims at maximizing the use of the available information in order to present with a satisfactory level of detail the influence of the systems engineering decisions. This approach is particularly useful when developing new processes since, in contrast to conventional LCA, the LCI cannot be generated from existing installations and statistical data. Optimizing process design and process size to incorporate the environmental criteria is extremely innovative and, considering it together with the supply chain and the associated geographic information, it is a breakthrough. In addition, Léda demonstrates how multi-objective optimization will generate

Pareto fronts that represent trade-offs between costs and environmental impacts while discussing the importance of considering several impact assessment indicators to finally select the best configurations.

The application on deep geothermal systems is adding three additional dimensions to the problem: the energy efficiency, the geology, and the uncertainty of dealing with renewable resources. The application of multi-objective optimization using three different objectives leads to trade-off maps that present the possible geothermal systems in a concise and synthetic way.

The chapter on the industrial ecology considers the development of urban concepts using process system design techniques. An interesting contribution is the use of life cycle inventory databases as simplified models for the system design. The use of mathematical programming techniques generates attractive scenarios for a complete community where the inhabitants'needs, the endogenous resources, and advanced technologies are all considered together in a holistic approach. The concept presented here already includes a lot of constraints to design more efficient communities, and there is a clear potential to extend the proposed models to more complex and more detailed communities with more numerous options. One could consider this approach as the ultimate urban energy system design tool to maximize the efficient usage of resources.

The work presented in this book is a clear contribution to the progress of sustainable system engineering and I hope that as a reader you will appreciate the results and enjoy the way Léda has presented them.

François Maréchal
Professor at EPFL

Preface

Our present civilization is rooted in high energy consumption. In order to maintain the same quality of life in Western countries and to improve it in developing countries, the world of tomorrow is likely to have an even greater need for energy. However, most of these needs are currently met by fossil energy sources, which are unsustainable on the long term, since their combustion emissions are the main contributors to a global warming threatening our society. Moreover, they have a limited availability. Therefore, cleaner and more renewable energy sources will have to be developed in the next decades to achieve sustainability. However, we must ensure that these new resources and technologies will not simply shift the impacts on the environment and create new ones. Many candidate technologies and resources currently compete or are still at the development stage in the field of renewable energy. Examples include the development of biofuels for mobility and deep geothermal resources for heat and electricity production. When assessing the environmental impacts of these new technologies, the full life cycle from cradle to grave must be accounted for, since the emissions associated with such systems often do not occur during utilization, but instead during the production or the disposal phases. Life Cycle Assessment (LCA) is a methodology that has been developed mainly in the last two decades to assess the environmental impacts of products, services or systems. However, it is important for emerging technologies using renewable energies that LCA is integrated at the conceptual design stage, since it is easier to take mitigation measures for impact reduction at this early stage than when the technology is already commercialized. In this way, LCA can really be used as a design tool and allow for identifying promising technologies that are also environmentally beneficial and economically competitive. This book proposes a multi-disciplinary approach to address the integration of LCA with process engineering techniques and provides application examples in the field of second generation biofuels and Enhanced Geothermal Systems. Moreover, it is becoming increasingly important to not only assess a given technology by itself, but also its integration at a larger scale and in more complex systems such as cities and territories. In such systems, multiple energy services have to be provided simultaneously and a given resource or technology is interacting with other resources and technologies. These interactions may concern competitions or synergies between the different potential resources and conversion pathways. This book addresses this aspect in the proposed approach and provides an application example within an urban energy system.

The work presented in this book is largely based on my PhD thesis, carried out at the Industrial Energy Systems Laboratory of the Swiss Federal Institute of Technology in Lausanne (EPFL) and funded by the Swiss Competence Center Environment and Sustainability. Conducting this work has been a very enriching experience both from a professional and personal perspective. I would like first to acknowledge my advisors, François Maréchal and Daniel Favrat, for their trust and the freedom they granted me in the realization of this research. Although I was initially hired to work on a project dealing specifically with geothermal energy, they were always very enthusiastic towards LCA and its integration in a broader approach that could be applied to other types of energy systems, such as biofuels or urban systems. Since I decided to work on a subject linked to various research fields that are very different from each other, I was fortunate to benefit from the expertise of some of my colleagues. In particular, I would like to thank Martin Gassner for our successful collaboration on biomass conversion, Samira Fazlollahi for her help with the models for urban systems, and Raffaele Bolliger for his support in implementing the LCA interface in our homemade software. In addition, I would like to thank all my colleagues who contributed to create the festive and professional atmosphere prevailing in the Industrial Energy Systems Laboratory, which was a great environment in which to conduct research. I am also grateful to all my friends for the good times we spent together during these years. Whether during dinners, parties, week-ends or trips that we shared together, you were always there to remind me how life is full of beautiful moments that are worth fully experiencing. I also went through some particularly difficult episodes during my thesis. For helping me through this, I would like to thank in particular my mother, Françoise, and my step-father, Pascal. Feeling understood and heard in such moments was a priceless and wonderful gift. Among the people who are as well dear to me, I would like to mention my brother, Brice, and my grand-parents, Nelly and Hubert. You five who represent my closest family have two invaluable qualities: the intelligence from the heart and the wisdom to live life simply. Thanks to you, I will never forget where I come from or who I am. Finally, there is a special place in my heart for my late father, Daniel, who instilled in me the curiosity and the desire to learn, two qualities that have been essential in completing this work.

Léda Gerber

Contents

CHAPTER 1

Introduction

During the last decades, the massive use of fossil energy resources to satisfy the needs in heating, electricity and transportation in developed countries has caused the release to the atmosphere of important quantities of previously sequestrated carbon. If the current trend continues, extending to the developing countries, humanity will face a global climate change in the next century. The perspective of mitigating this threat and the awareness of the limited availability of fossil resources has lead to a growing interest in developing large-scale alternatives based on renewable energy resources.

In Switzerland, this need has even been strengthened since the recent decision of the Federal Council to abandon nuclear power by 2034. As defined in its national energy strategy for 2050, the new energy policy is based on the concept of the 2000W society and 1 ton of emitted CO_2-equivalent per inhabitant and per year (Swiss Federal Council (2011)). A 2000W society means that the yearly energy consumption per capita, including all the sources of energy necessary to supply the required services such as heating, electricity and transportation, should not be higher than 2000W when this consumption is calculated as an average power requirement. However, as illustrated by Figure 1.1, the current energy consumption per capita in Switzerland is far from this objective, with a primary energy requirement of around 4600W per inhabitant. It remains mostly based on fossil energies, which results in around 7 tons of CO_2-equivalent per inhabitant and per year (OFEV (2011)).

In order to reach these objectives, the Swiss energy policy targets simultaneously an increase in the energy efficiency and in the substitution of fossil energies and nuclear by renewables, still representing a minor share of the total energy consumption, except for hydro-electricity. Thus, the potential for geothermal energy, photovoltaics, biomass and wind energy remains nowadays largely underexploited. In the future, several emerging technologies could contribute to increase particularly the share of geothermal energy and of biomass in an efficient way. In particular, the combined production of electricity and

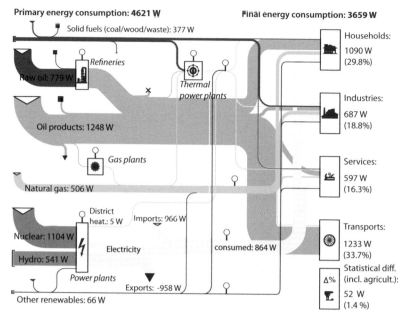

Figure 1.1 Detailed energy flowchart of Switzerland in 2010, in W per capita (adapted from OFEN (2010b)).

heating by the exploitation of deep geothermal resources as Enhanced Geothermal Systems (Minder et al. (2007)) or the production of synthetic natural gas by the thermochemical conversion of biomass (Stucki et al. (2010)) appear to be promising. Nevertheless, such technologies are for the moment still at the stage of pilot-scale projects (Genter et al. (2012); Stucki et al. (2010)) and several technological developments are necessary to make them economically competitive.

However, the preliminary evaluation of such technologies before they enter the market should not be limited to economic and efficiency criteria. Indeed, it should be ensured that their overall environmental balance is beneficial not only with respect to their fossil competitors, in terms of greenhouse gases emissions but also for other environmental impacts, such as eutrophication, acidification, toxicity or land use aspects, for example. To conduct such an assessment, it is necessary to use a so-called 'life-cycle' based approach, since the major impacts from renewables may often not only occur during the use phase, but also during the production phase – the 'cradle' – or the disposal phase – the 'grave'. An illustrative example of this necessity is the production of biofuels from several feedstocks and using different potential processes to replace fossil fuels. The graph on the left of Figure 1.2, adapted from Zah et al. (2007), shows that, when compared with their fossil competitors on a life-cycle basis integrating not only the fuel combustion in the engine, but also the feedstock cultivation, the biofuel production and its transport, some of the candidate feedstocks and processes do not result in an important reduction in the greenhouse gases emissions per unit of mobility provided, here the person-kilometer. Moreover, the overall environmental performance appears to be even worse for some of the biofuels than for fossil fuels when other types of impacts are considered, as illustrated by the graph on the right of Figure 1.2. In this example, the environmental impacts considered are the effects on the human health, on the ecosystem quality and on the resources and are combined together in a single score of performance, the Ecoindicator99 (Goedkoop and Spriensma (2000)).

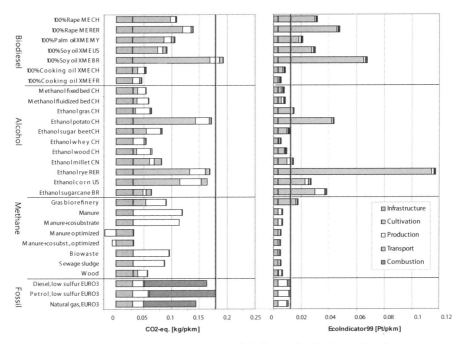

Figure 1.2 Environmental comparison of different biofuels production processes and feedstocks with fossil petrol as the reference (red line), in terms of greenhouse gases emissions and total environmental performance calculated with Ecoindicator99 (adapted from Zah et al. (2007)).

Therefore, evaluating the environmental life-cycle performance of candidate technologies in the field of renewable energies is crucial before they enter the market, in order to avoid to cause large-scale environmental or social impacts, like this has been the case for first generation biofuels. Furthermore, mitigating the environmental impact is easier and less costly at an early stage in the process development. Eventually, this allows for integrating environmental criteria along with economic and thermodynamic ones and to relate these three types of criteria to each other, in order to take the best decisions for the final commercial design of the mature technology, which should respect the principles of economic, environmental and social sustainability.

In order to assess the environmental performance of such emerging technologies in the field of renewable energy, Life Cycle Assessment (LCA) is currently used as the state-of-the art method. Also sometimes termed as Life Cycle Analysis,

LCA is a quantitative methodology for assessing the environmental impacts of a product, a service, or a system, related to its function and accounting for its overall life cycle.

The methodology, which scientific bases were published in the 1990s, was originally intended for product development and comparison in the field of environmental management (Guinée et al. (1993a,b)). At that time, the LCA methodology was not yet fully established, which led to a number of critics (Guinée et al. (1993c); Ayres (1993); Finnveden (2000)), notably about arbitrary choices that could influence the outcome of such studies. This eventually resulted in the establishment of the ISO norms 14040 & 14044 to provide a fixed methodological framework and guidelines to conduct a LCA (ISO (2006a,b)). However, LCA is still a relatively new field of research needing several developments related to a number of issues (Finnveden et al. (2009)), including, among others, the definition of scenarios for future systems in consequential LCA, the allocation issues when dealing with multiple input, multiple output and recycling processes and the use of databases representing average market technologies, like ecoinvent® (Frischknecht et al. (2005)).

Conventional LCA has been widely applied to the quantification and analysis of environmental impacts from renewable energy systems. The example of LCA studies applied to biofuels production processes (Kaltschmitt et al. (1997); von Blottnitz and Curran (2007); Zah et al. (2007)) highlights the undesirable impacts caused by the production of first generation biofuels. This justifies to use such an approach in the case of renewable energy systems to account for life cycle impacts. Moreover, these examples show the importance of accounting for other environmental impacts – e.g. acidification or eutrophication – than the global warming potential, which is often the only environmental impact considered outside of the community of environmental analysis. Such studies consider the technologies that are available on the market and sometimes already used for large-scale production, such as bio-ethanol production.

However, as already stated in 1993 by Keoleian, more freedom is available in the design at the conceptual stage of a product or of a technology. This highlights the importance of an early assessment of emerging technologies and makes therefore easier the integration of measures for impact mitigation. A major drawback is that at this early development stage, establishing a detailed life cycle inventory (LCI) is more difficult. The conventional approach of environmental analysis to this issue is to base the inventories on pilot- or lab-scale technologies. Among others, examples exist in the field of biofuels, such as the evaluation of environmental impacts caused by the Synthetic Natural Gas (SNG) production by thermochemical conversion of ligno-cellulosic biomass (Felder and Dones (2007)), or in the field of geothermal systems, for the electricity production from a low-temperature Enhanced Geothermal System (EGS) (Frick et al. (2010)). Though such studies allow for the identification of critical inventory elements and environmental hot spots, they do not reflect the final commercial-scale design of such technologies and do not account for the engineering decision variables, such as the process operating conditions, or for the technological alternatives, such as the type of cycle used for power production. Pehnt (2006) and Steubing et al. (2011b) address the issues of defining future scenarios and of performing sensitivity analyses by accounting for changes such as process efficiency and background electricity mix, using average static values taken from the literature. Therefore, though this type of approach reflects better an average technological evolution, it can hardly be integrated in the final decision-making at the conceptual stage with other economic and thermodynamic criteria.

The LCA of emerging renewable energy technologies does not only concern the technology itself, but also its interactions with other components, included or not in the system boundaries. One of these interactions concern the supply chains of background resources or materials. This has been addressed by Pehnt (2006), who considers an evolution of the background processes with respect to time, though without considering the possibility of taking decisions on the supply chain itself. Another interaction to be considered is the substitution of energy services currently produced by fossil energy-based technologies. Steubing et al. (2012) discuss the optimal services and technologies that have to be produced from biomass feedstocks when substituting fossil energy, though considering static models of average market technologies and without accounting for the other renewable competitors.

Several authors worked on the extension of the LCA methodology to include energy and economic aspects. Cornelissen and Hirs (2002) propose to combine LCA and exergy analysis to analyze the exergy flows involved in a complete life cycle. Exergy refers to the quality of an energy flow and represents the potential of a system to produce useful work. Cornelissen and Hirs demonstrate that this indicator is appropriate to assess the depletion of natural resources. Regarding economic aspects, Life Cycle Costing is a methodology developed of its own, similar to the environmental LCA and differing from the traditional economic analysis by accounting for the external and hidden costs of a product (Shapiro (2001, 2003)). Finally, Guinée et al. (2011) discuss the development of a Life Cycle Sustainability Analysis framework, which would broaden the scope of LCA by including the other sustainability criteria – i.e. economic and social. Such a framework would therefore integrate a collection of trans-disciplinary models.

Analyzing the available literature specifically in the field of LCA, it appears that the environmental evaluation of emerging technologies for renewable energy conversion and that the integration of the environmental dimension in a broader sustainability framework are presently recognized as topics of importance in the LCA community. Several studies have been conducted on the early assessment of renewable energy technologies, and strategies have been developed to model the changes in the technology evolution, in order to integrate the environmental aspects in decision-making. However, there is no study about the advantages of linking the LCA of such technologies with the results of a process designed by the engineers, which would represent a real-scale commercial technology.

When it comes to commercial-scale technologies, the community of process systems engineering has developed several efficient methods for the optimal design and synthesis of industrial processes and energy systems. Such methods include thermo-economic analysis, which allows for relating process operating conditions and design to costs (Frangopoulos (1987)), structural optimization of process flowsheets, which allows for selecting the best technological alternatives for a given process, (Papoulias and Grossmann (1983a,b,c)) and process integration techniques, which allows for accounting for the potential internal heat and power recovery within a given process (Linnhoff et al. (1982); Floudas et al. (1986); Maréchal and Kalitventzeff (1998); Kemp (2007)). These methods have been widely applied for costs minimization and for increasing the thermodynamic efficiencies in the synthesis of industrial processes and in the conceptual design of energy systems.

Several other authors in this field worked on the development of a methodology to integrate LCA and indicators of environmental performance in process systems design, synthesis and optimization. The first works were conducted by Azapagic and Clift (1995) and Stefanis et al. (1995), who proposed the initial methodologies to integrate the cumulated environmental impacts as a criterion for the operation of chemical processes. The following studies of Kniel et al. (1996), Azapagic and Clift (1999) and Alexander et al. (2000) deal with the use of LCA in a multi-objective optimization framework to calculate the trade-offs existing between economic and environmental objectives. Again, these studies are specifically applicable to the context of chemical processes operation.

With the evolution of process design and optimization techniques, new methodologies have been proposed to integrate the LCA in a more robust process design framework. Hugo and Pistikopoulos (2005) introduced a strategy to synthesize supply chains, considering both costs and cumulated life cycle impacts in a multi-objective optimization framework using mixed integer linear programming (MILP) techniques. Sugiyama et al. (2006) proposed as well a similar strategy for the selection of technological alternatives. Later, these approaches were extended to the extraction of optimal flowsheets configurations from the process superstructure, which represents all the potential technologies that can be selected or not to design a process, and to mixed integer non-linear programming (MINLP) problems (Guillén-Gosálbez et al. (2008)). The uncertainties in the supply chain synthesis were as well included later in this approach by Guillén-Gosálbez and Grossmann (2010). Another emerging approach for the minimization of the environmental impacts and costs is the application of industrial ecology to process systems design. Industrial ecology aims at minimizing the use of resources and subsequently the environmental impacts by identifying the possible energy and material exchanges within a considered system (Allenby and Richards (1994); Ehrenfeld (1997); Erkman (1997)). The integration of these principles has been realized for industrial processes by several authors (Urban et al. (2010); Diwekar and Shastri (2010); Cimren et al. (2012)), though their studies do not consider the application of LCA for calculating the environmental performance.

The above studies are all conducted in the field of chemical processes, and consider thus only economic and environmental criteria, without the thermo-dynamic dimension, such as the energy or the exergy efficiencies (Borel and Favrat (2005)), which are of crucial importance in the field of energy systems analysis. von Spakovsky and Frangopoulos (1993a,b) introduced the concept of environomics for energy systems by taking into account not only the total costs, but also the exergy and some environmental aspects, such as direct emis-sions and resource consumption, but without specifically referring to LCA. This type of approach was then applied to the design of district heating systems (Li et al. (2004)), cogeneration plants (Lazzaretto and Toffolo (2004)) and com-bined cycle power plants with CO_2 separation options (Li et al. (2006)) using multi-objective optimization techniques. Later, Papandreou and Shang (2008) and Martinez and Eliceche (2009) introduced the use of LCA in process design framework for the minimization of CO_2 emissions from fossil power plants, however without considering process integration and focusing on on-site emis-sions. The integration of LCA and of process integration has been addressed by Luterbacher et al. (2009) for the production of Synthetic Natural Gas (SNG) by hydrothermal gasification. Eventually, Bernier et al. (2010) combined process integration, multi-objective optimization techniques and LCA for the design of natural gas combined cycles (NGCC) power plants considering CO_2 capture options.

Emphasizing on environomic energy systems design and analysis, all the above studies do not consider systems based on renewables but on fossil resources, except for Luterbacher et al. (2009), which considers biomass resources. How-ever, thermo-economic analysis and optimization techniques have been already successfully applied to the conceptual design of renewable energy systems. Ex-amples include the design of thermochemical processes for the conversion of biomass to SNG (Gassner and Maréchal (2009a,b)) and liquid biofuels (Tock et al. (2010)), as well as the design of organic Rankine cycles (ORC) for power production from geothermal resources (Lazzaretto et al. (2011)).

From this literature review in the field of process systems design, it appears that several strategies have been developed for the integration of LCA in frame-works for process design and optimization, in particular for chemical processes and fossil energy systems. None of them specifically considers the application to the conceptual design of renewable energy systems. Though it is generally agreed that the flows of the life cycle inventory have to be linked to the process flowsheet, no specific methodology has yet been reported for identifying these flows according to the LCA system boundaries and for establishing the mathe-matical formulations that allow for linking them to the process superstructure or flowsheet. Moreover, Grossmann and Guillén-Gosálbez (2010) have recently observed that one of the major remaining limitations of the application of LCA methodology to process systems design is the lack of a systematic method for generating and identifying process alternatives that minimize the life cycle impact while still yielding good economic performance. Furthermore, none of these existing methods has demonstrated clearly the advantage of using the LCA in conjunction with process integration techniques compared to a conventional LCA that is usually conducted by the environmental analysts for the evaluation of emerging technologies. Another important aspect is that most of the studies mentioned above are process-design oriented. Hence, they do not discuss the issues that may be specific to the LCA methodology, such as the influence of the system boundaries, the choice of functional unit or the allocation of by-products, which are often mentioned as problematic issues of the LCA methodology in the literature specific to the LCA field (Finnveden et al. (2009)).

Comparing both the fields of environmental analysis and process systems design, it appears that several studies have been conducted over the last 20 years on the LCA and its application to the conceptual design of renewable energy conversion technologies, but that they do not focus on the same aspects.

The studies conducted by the environmental analysts for the application of LCA to renewable energy technologies follow the strict and detailed framework established by the ISO-norm for the LCA methodology: the function and system boundaries are clearly defined, databases are generally used to account for off-site emissions of auxiliary materials and impacts due to construction and end-of-life, and several works treat of the allocation or substitution issues when dealing with multiple outputs or inputs, or recycling loops. There is as well an interest in accounting for the competition between different potential candidate technologies or uses of a given resource. However, when it comes to the evaluation of future or emerging technologies in the field of renewable energy, average values from the literature are used to introduce changes in the system by defining *a priori* a few scenarios. Hence, no detailed model of the technology is used to calculate the inventories, which would allow for a dynamic interaction between the design and the impacts. Moreover, though there is a growing interest in integrating LCA in a more global framework for sustainability evaluation and that life-cycle based methods have been developed for costing and exergy analysis, tools to calculate the trade-offs between these conflicting objectives, such as multi-objective optimization techniques, have not yet been used. It makes thus preliminary decision-making difficult for the evaluation of the potential designs at the conceptual stage.

On the other hand, the methodologies developed by process systems engineers rely on robust tools and frameworks for process systems design, such as process integration, thermo-economic modeling and multi-objective optimization algorithms, which allow for calculating dynamically the trade-offs between conflicting objectives. Moreover, they generally consider in a systematic manner the decision variables regarding process operation, design and technological alternatives embedded in the superstructure. In the field of industrial processes, methodologies have been already developed to synthesize automatically the best supply chains accounting for environomic criteria and to identify the industrial ecology possibilities. However, the integration of environmental impacts in the process design procedure has been mostly limited to the design of fossil energy systems focusing on on-site emissions, and the application of LCA to the design of chemical processes. Thus, the specificities linked with the environmental aspects of the conceptual design of renewable energy technologies are not addressed. These concern mainly life-cycle aspects, off-site emissions from auxiliary materials, a broad range of environmental impacts to be considered, and allocation or substitution issues, these ones being crucial in the case of a technology producing multiple energy services.

Therefore, both fields have used different approaches for the sustainable design of renewable energy conversion systems and the decision-making related to it. The synthesis of the best conversion chains, or supply chains, is as well a domain of interest in both fields. While each one of these two distinct approaches has its advantages, they also both suffer from several lacks and weaknesses. However, when they are compared with each other, it can be pointed out that the strengths of one approach generally respond to the weaknesses of the other. Thus, the gap between the two fields can be bridged by bringing the environmental analysis and the process systems engineering to work together in a common framework for the conceptual design of renewable energy technologies. Such a framework should combine in a single methodology process integration, thermo-economic modeling and analysis, multi-objective optimization and Life Cycle Assessment.

The objective of this book is precisely to present in a first time such a systematic methodology for the conceptual design of renewable energy conversion systems, by integrating the LCA in a process systems design framework. Its second objective is to present the application of the developed methodology to several case studies illustrating the different aspects included, in order to show that the integration of LCA with process systems design can be used to orientate the decision-making related to the future development of renewable energy conversion systems.

CHAPTER 2

The LCA approach

2.1 Computational framework

The original computational framework used to implement the methodology aims at structuring and organizing the information in order to represent the possible interactions between the different components to be considered in the energy system design (Gassner and Maréchal (2009a) and Bolliger (2010)) . The optimization problem aims at calculating the configurations of the system that minimize simultaneously the non-linear costs, impacts or maximize the thermodynamic efficiencies. It is therefore by essence a Mixed Integer Non-Linear Programming (MINLP) multi-objective optimization (MOO) problem. It is solved following a two-stage decomposition methodology with a master problem and a slave sub-problem, and is described in Figure 2.1, with the original parts displayed in black.

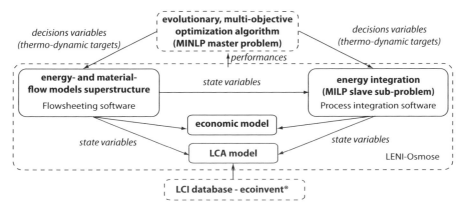

Figure 2.1 Computational framework for system simulation and design (adapted from Gerber et al. (2011a)).

A superstructure including the different technological options is built and the thermo-economic models of these components are developed. In a first time, flowsheeting software is used to calculate the energy and mass flows for the different models, for a given set of operating conditions. The resulting energy flows are then used in a second time to generate the process integration model, which optimizes the heat recovery and the combined fuel, heat or power production, depending on the problem to be solved and on the considered technology. This is done by minimizing the operating cost while computing the energy integration (Linnhoff et al. (1982); Maréchal and Kalitventzeff (1998)). The process integration model is solved as a Mixed Integer Linear Programming (MILP) sub-problem, which decision variables are the utilization rates of the different technologies in the superstructure. It is submitted to the constraints of the heat cascade. Eventually, the thermodynamic states and flow rates of the energy-flow and energy integration models are used in a post-calculation phase to size the equipment, estimate the cost and evaluate the performances of the process configuration. Combining the operating performances with the investment allows for developing economic and thermodynamic indicators that can be used at a master level to realize a multi-objective optimization. A set of decision variables addressing technology choices and operating conditions of the process and utility system is provided for this. The multi-objective MINLP master problem is then solved using an evolutionary algorithm (Molyneaux et al. (2010)).

In its original form, the performance evaluation is limited to an economic model that rates the equipment in order to meet the thermodynamic design targets (Gassner and Maréchal (2009a)). This computational framework is extended to systematically include the environmental life-cycle performance indicators with respect to the detailed process design in the performance evaluation. In analogy with the economic model, it exploits the flowsheeting and the energy integration results – i.e. material and energy flows, equipment sizes – to calculate the life cycle inventory (LCI) of emissions and extraction flows of single substances associated to the process equipment and its operation. It is illustrated by the red part on Figure 2.1. The LCI is based on reference data in EcoSpold format (Frischknecht and Jungbluth (2007)) from the ecoinvent® life cycle inventories database version 2.1 (Frischknecht et al. (2005)). Finally, a life cycle impact assessment (LCIA) calculation is performed for the obtained LCI. The impact categories from the LCIA phase are used as indicators of the environmental performances of the process configuration, and can be considered in its multi-objective, environomic – i.e. energetic, environmental and economic – optimization.

Section 2.2 discusses in details the integration of the LCA methodology within the computational framework.

2.2 LCA methodology integration

2.2.1 LCA basics

As stated by the ISO norms, the LCA consists in four mandatory stages: the goal and scope definition, the life cycle inventory, the impact assessment and the interpretation, this fourth stage being as well necessary during all the other stages of the LCA. During the goal and scope definition, the objectives of the LCA are stated, the boundaries of the considered system are set, and the functional unit (FU) is defined. This functional unit is related to the function of the system and represents the reference quantity for every element considered in the system or crossing its boundaries. In the second stage, the life cycle inventory (LCI), all the emissions and extractions of single substances involved in the life cycle of the considered system are quantified and summed in a vector of substances, considering the different elements of the LCI:

$$Em_{j,i}^{FU} = \frac{Em_{j,i}}{FU_{tot}} \tag{2.1}$$

where $Em_{j,i}$ is the emission of the elementary flow i for the LCI element j and FU_{tot} is the total functional unit quantity involved in the life cycle of the considered system. The total quantity of an emission or an extraction is then calculated by:

$$\forall i = 1...n_i : \ Em_i^{FU} = \sum_{j=1}^{n_j} Em_{j,i}^{FU} \tag{2.2}$$

where n_j is the number of LCI elements.

In the third stage, the impact assessment, these single substances are aggregated in a reduced number of impact categories – midpoint or endpoint – having an environmental significance. This is done by using a matrix, termed as an impact assessment method, which contains the weightings for all the substances considered in the LCI:

$$\begin{bmatrix} F_{1,1} & ... & F_{1,n_i} \\ ... & ... & ... \\ ... & F_{l,i} & ... \\ ... & ... & ... \\ F_{n_l,1} & ... & F_{n_l,n_i} \end{bmatrix} \cdot \begin{bmatrix} Em_1^{FU} \\ ... \\ Em_i^{FU} \\ ... \\ Em_{n_i}^{FU} \end{bmatrix} = \begin{bmatrix} I_1^{FU} \\ ... \\ I_l^{FU} \\ ... \\ I_{n_l}^{FU} \end{bmatrix} \tag{2.3}$$

where $F_{l,i}$ is the weighting factor to convert the LCI emission i into the impact category l, Em_i^{FU} is the emission or extraction i calculated in the LCI, and I_l^{FU} is the impact category l of the impact assessment method. Some of the impact

assessment methods propose an additional weighting step by aggregating the
impact categories into a final single score:

$$I_{tot}^{FU} = \sum_{l=1}^{n_l} I_l^{FU} \cdot w_l \qquad (2.4)$$

where w_l is the factor used for the normalization and weighting of the different
impact categories.

2.2.2 Guidelines for integration in system design

The developed general methodology for the construction and the integration
of the LCA model within a thermo-economic model complying with the ISO
norms is depicted in Figure 2.2. It comprises the four mandatory stages of
a LCA. The parts of particular importance to link the LCA model with the
process design and configuration are displayed in black. These developments
in the methodology concern mainly the Life Cycle Inventory stage and are
discussed in details in subsection 2.2.3.

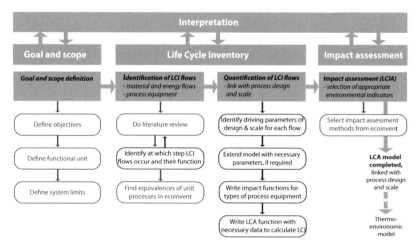

Figure 2.2 Developed general methodology for LCA model integration in process
design framework (adapted from Gerber et al. (2011a)).

While listing the LCI flows through the system boundaries, it is crucial to
identify to which process units of the superstructure the flows are linked. This
is necessary because they are then mathematically linked with the thermo-
economic model. The scaling of impacts due to the changes in operating conditions and to the size of the process equipment is therefore accounted for. A
LCA function including all the LCI flows and the impact due to process equipment allows for calculating the whole LCI of a given process configuration.
The ecoinvent® life cycle inventories database is used to find equivalences for
all LCI elements and impact assessment methods. The use of such a database
allows to account for the induced off-site emissions.

2.2.3   Life Cycle Inventory elements

The defined system is taken as a basis to identify the different material and energy flows of the inventory. Their associated life-cycle elementary flows of emissions and extractions are then added to each other. Since the LCI database proposes aggregated datasets for unit processes, it is not necessary to calculate individually each single emission or extraction for all the LCI elements. The emissions and extractions vector of the elementary flows for each LCI element is calculated by:

$$Em_{j,i}^{FU} = em_i \cdot V_j^{FU} \qquad (2.5)$$

where $Em_{j,i}^{FU}$ is the emission or extraction of substance i for the LCI element j, em_i is the specific emission or extraction per unit of LCI element, taken from the LCI database and V_j^{FU} is the quantity of the LCI element j per functional unit (FU). For the single emissions or extractions that have to be included as such in the LCI, the amount of the flow $Em_{j,i}$ is directly calculated.

To establish the LCI, three different categories of LCI elements are distinguished in the present methodology:

1) Flows of the thermo-economic model. These are the flows related to the process operation that are directly identified on the process flowsheet, from the process integration results or from the calculations of the economic model. Examples of such elements are the consumed or the produced electricity, or the water required for a chemical reaction. In the case of the produced energy services, since the present methodology is applicable for renewable energy systems, the avoided impacts from the produced energy services are calculated by substitution of the equivalent services produced from fossil sources. Moreover, the substitution allows to account for the conversion efficiency of the technology.

2) Flows of auxiliary materials and emissions. These are the flows that are required to support process operation and that are not included in the process flowsheet or in the economic model, but that have nevertheless an environmental significance. It includes the process auxiliary materials, the emissions not calculated by the process flowsheet, the waste that have to be disposed off or the logistics. Examples of such elements are the auxiliary combustion bed materials or the particulate matter emissions from a reactor.

3) Process equipment. This is the equipment necessary to operate the process. Examples are the reactors or the heat exchangers.

Figure 2.3 illustrates the extension of a thermo-economic model for a renewable energy system to a LCA model with these different LCI elements that have to be accounted for. For the substitution of energy services, though the quantities of energy services supplied by the technology are calculated by the thermo-economic model, the assumptions made for the substitution depend on the LCA model.

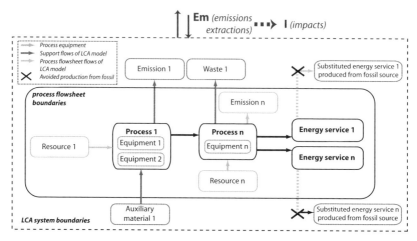

Figure 2.3 Types of elements accounted for in the LCI when extending the thermo-economic model to a LCA model.

Details about the identification and quantification of these three categories are given below. Examples of application of the methodology for LCI scaling are provided for biofuels production processes at Chapter 3, and for geothermal energy conversion systems at Chapter 4.

Flows of the process flowsheet

In this particular case, it is not necessary to develop a mathematical expression to calculate the quantities of these LCI flows. These values are already calculated as flow rates in the process flowsheet model used for the thermo-economic calculations and are directly used for the LCA. Then, they are expressed per FU using the following equation:

$$V_j^{FU} = \frac{\dot{V}_j(x_d) \cdot t_{yr} \cdot r_o}{FU_{tot}(x_d)} \tag{2.6}$$

where $\dot{V}$ is the flow rate of the LCI flow j taken from the process flowsheet and varying with the decision variables of the master optimization problem x_d, t_{yr} is the installation lifetime, r_o is the yearly percentage of operation of the installation, and FU_{tot} is the total FU quantity involved in the life cycle of the studied system, calculated by:

$$FU_{tot} = \dot{FU}(x_d) \cdot t_{yr} \cdot r_o \qquad (2.7)$$

where $\dot{FU}$ is the flow rate of functional unit calculated by the process flowsheet, as well function of the decision variables x_d .

Flows of auxiliary materials

These flows are not included in the original process flowsheet but are required for the conversion and are therefore indirectly related to it. In this case, it is necessary to develop models to calculate their quantities. These mathematical expressions have to be based on the values calculated in the process flowsheet, and the amounts are therefore indirectly linked with it. A case-by-case approach has to be adopted in order to identify to which parameters of the process flowsheet these flows are linked. However, the parameters that have an influence on the amount of auxiliary materials for the operation phase, termed more often as the use phase in the LCA community, can be expressed as:

$$\dot{V}_{j,o}^{FU} \sim V_{j,init}(x_d), \dot{\alpha}_j, t_{yr}, r_o \qquad (2.8)$$

where $V_{j,o}^{FU}$ is the flow rate of auxiliary material j, $V_{j,init}$ is its initial quantity, function of the decision variables x_d, and $\dot{\alpha}$ is the turnover of the material, in quantity of j per second.

For the support materials depending on the construction or on the end-of-life phase, the formulation simply becomes:

$$V_{j,c}^{FU}, V_{j,e}^{FU} \sim V_{j,init}(x_d) \qquad (2.9)$$

The quantities $V_{j,init}$ have to be calculated from the literature or as a function of the decision variables.

Process equipment

An inventory of all major process equipment is first set up, and a specific type is assigned to each unit. For consistency with the thermo-economic model, the inventory is based on the equipment that is rated and quoted in the economic model. For each type of process equipment, equivalent LCI elements are found in the LCI database. If no equivalence is found, the quantity of materials required for the construction of the process equipment and its transport needs to be estimated. The equivalent LCI contribution is thus deduced from the corresponding materials and transports.

The goal is to calculate the life cycle inventory of each process equipment considering its size and operating conditions. Such inventory functions can be based on existing inventories of similar equipment and have to take the process conditions into account. Prior to establishing the methodology for the impact scaling of process equipment, it is necessary to define the subsystems limits to be considered for the LCA of the process equipment. These limits are displayed in Figure 2.4. Recycling is not included, since the flows are assumed to leave the system.

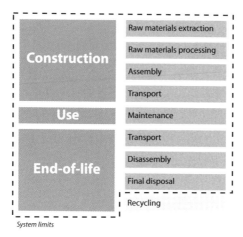

Figure 2.4 Systems limits for LCA of process equipment (adapted from Gerber et al. (2011a)).

The developed impact scaling method for process equipment is presented below.

2.2.4   Impact scaling method for process equipment

In classical LCA, emissions and impacts of the process equipment are most often linearly scaled with equipment size, as for example in Felder and Dones (2007). This assumption is yet not justifiable in general, since the emissions are likely to be proportional to the amount of materials required to produce the process equipment. This amount is generally not linear with respect to the size of the process equipment, since a smaller size is likely to require a higher input of materials per unit of capacity than a larger size. It is thus better suited to scale the emissions related to the process equipment with a more general form that is similar to the formulation for equipment costs estimation, where a specific exponent is used for each single emission, i.e.:

$$\frac{Em_{j,i}}{Em_{j,ref,i}} = n \cdot \left(\frac{A_j(x_d)}{n \cdot A_{j,ref}}\right)^{k_{j,i}} \cdot c_j, A_j \in [A_{j,min}; A_{j,max}] \tag{2.10}$$

$$n = [int(\frac{A_j}{A_{j,max}}) + 1] \tag{2.11}$$

where $Em_{j,i}$ is the scaled emission of the elementary flow i, $Em_{j,ref,i}$ the emission of the reference LCI dataset, A_j the functional parameter, or the capacity, related to the size of the process equipment j in the validated range $[A_{j,min}; A_{j,max}]$, varying with the decision variables x_d, n the number of units required in parallel, $A_{j,ref}$ the value of this functional parameter for the reference dataset, $k_{j,i}$ the scaling exponent of the elementary flow i and c_j a correction factor that represents the specific operating conditions or the unit type. The formulation of Equation 2.10 reduces to classical linear scaling by setting $n = 1$, $k_{j,i} = 1$, $c_j = 1$, but also allows for taking into account the effects of positive and negative economies of scale on the environmental impacts ($n < 1$ and $n > 1$, respectively).

The scaling is done for each single emission or extraction of the LCI, and not on the final impacts in order to track or analyze single substances. Finally, these emissions have also to be expressed per FU using Equation (2.1).

The general methodology for identifying suitable parameters for Equation (2.10) is summarized step-by-step in Figure 2.5.

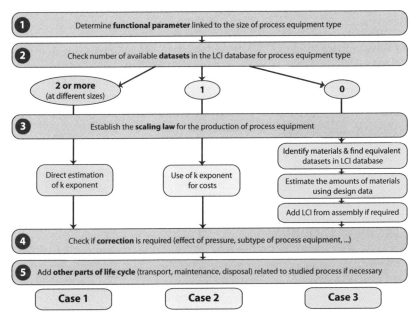

Figure 2.5 Methodology to establish scaling laws for process equipment (adapted from Gerber et al. (2011a)).

The calculation of the impacts for the production phase, excluding the transportation of the manufactured process equipment to its place of use, has been outlined in Equation (2.10). The impact scaling laws definition concerns first the determination of the functional parameter A that represents the size of the type of process equipment. Then, the exponents $k_{j,i}$ of Equation (2.10) have to be estimated. This estimation is however subject to the availability of data from the LCI database. Indeed, the direct determination of the exponents $k_{j,i}$ for each emission of the LCI involves that two or more reference datasets at different sizes for the same type of process equipment are available from the LCI database. However, current LCI databases generally do not contain enough information on the production of process equipment. Therefore, other strategies to estimate the exponents $k_{j,i}$ have been developed in the case where not enough data are available from the LCI database.

Scaling methods

Three different cases are distinguished, and for each case a method is defined to establish the impact scaling law from the process equipment.

Case 1. At least two datasets are available in the LCI database for the type of process equipment. In this case, the exponents $k_{j,i}$ can directly be determined for each LCI emission, transforming Equation (2.10) into:

$$k_{j,i} = \frac{logEm_{j_1,i} - logEm_{j_2,i}}{logA_{j_1} - logA_{j_2}} - logc_j^* \qquad (2.12)$$

where $Em_{j_1,i}$ is the emission i for the reference dataset with the value A_{j_1} for the functional parameter linked to the size, and $Em_{j_2,i}$ is the emission i for the reference dataset with the value A_{j_2} for the functional parameter, and c_j^* is a constant related to the correction factor c_j. If more than two datasets are available, a linear regression is performed on the following equation:

$$log\left(\frac{Em_{j,i}}{Em_{j_{ref}}}\right) = k_{j,i} \cdot log\frac{A_j}{A_{j,ref}} + logc_j \qquad (2.13)$$

Case 2. Only one dataset is available in the LCI database for the type of process equipment. In this case, the exponents $k_{j,i}$ can not be directly determined for each LCI emission. By similarity between the economic and the environmental scaling laws, it is then assumed that the ratio of the costs is equal to the ratio of the emissions at two different sizes. Therefore, the emission exponents $k_{j,i}$ of the LCI emissions are assumed to be equal to the one for the investment cost scaling, and the impact scaling law becomes:

$$\frac{Em_{j,i}}{Em_{j,i,ref}} = \frac{CI_j}{CI_{j,ref}} \qquad (2.14)$$

where $Em_{j,i}$ is the scaled emission of the elementary flow i, $Em_{j,i,ref}$ is the reference emission of the LCI dataset, CI_j is the scaled investment cost, and $CI_{j,ref}$ is the reference cost. CI_j and $CI_{j,ref}$ are both calculated using the well-established correlations from Ulrich (1996) and Turton et al. (1998).

Case 3. No dataset is available in the LCI database for the type of process equipment. In this case, it is necessary to perform the LCA of the process equipment by considering the different materials of construction. Equivalences are then found in the LCI database for these materials. A scaling law has then to be established to calculate the quantity of each one of these materials, using design data at different sizes. The scaled emissions for the considered type of process equipment become:

$$Em_{j,i} = \sum_{m=1}^{n_m} Em_{j,i,m} \qquad (2.15)$$

where $Em_{j,i}$ is the scaled emission of the elementary flow i, $Em_{j,i,m}$ is the scaled emission of the elementary flow i from the material m composing the process equipment j, and n_m is the number of construction materials for the process equipment j.

As shown by Figure 2.4, it is necessary to include not only the production of process equipment, but also the use phase including maintenance, and the end-of-life phase including disassembly, transportation and disposal.

By analogy with the economic assumptions, where maintenance is commonly assumed to represent 5% of the costs per year, maintenance is assumed to represent 5% of the total impact of the process equipment per year of operation. To account for transportation in the production, the end-of-life stages and the disposal, the mass of process equipment is calculated. It is assumed that the mass follows a similar law to the cost and impact scaling laws, since it is as well proportional to the amount of materials required to build the process equipment:

$$\frac{m_j}{m_{j,ref}} = (\frac{A_j}{A_{j,ref}})^{k_{j,m}} \qquad (2.16)$$

where m_j is the mass of the process equipment j to be calculated, $m_{j,ref}$ is the mass of a reference process equipment, A_j is the functional parameter of the process equipment j and $k_{j,m}$ is an exponent.

Relevance of the approach

The relevance of this approach for the impact scaling of the process equipment is illustrated in Figure 2.6(a) and 2.6(b) by two examples, a shell-and-tube heat exchanger and a compressor. For the heat exchanger, no LCIA data are available at two different sizes or more from the LCI database. Detailed design data are taken from a heat exchanger manufactured for an industrial chemical process at 10 different sizes, the size being represented by the exchange area in m^2. It is composed by two materials, stainless steel and unalloyed steel, which quantities are calculated for each exchange area using the design data. Then the resulting impact is calculated using the LCIA data for the two materials from the LCI database, using the single score of the Ecoindicator99-(h,a) (Goedkoop and Spriensma (2000)), which is an impact assessment method that proposes a combined single indicator for damages caused on human health, ecosystem quality and non-renewable resources. The results per unit of exchange area are displayed in Figure 2.6(a). A comparison of the impact calculated by the design data with a linear extrapolation and with a power extrapolation using a costs exponent from the literature (Turton et al. (1998); Ulrich (1996)) for one of the design points at 13 m^2 is as well presented.

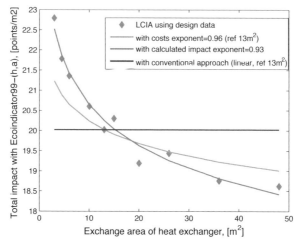

(a) Impact scaling for a shell-and-tube heat exchanger, using costing analogy and linear extrapolation for one point, and comparison with LCIA conducted using design data (Case 3).

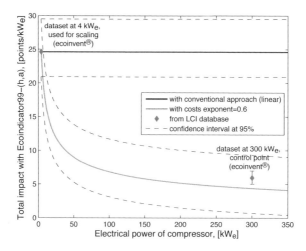

(b) Impact scaling for a compressor, using costing analogy and linear extrapolation, and comparison with LCIA data from ecoinvent® (Case 1).

Figure 2.6 Examples of impact scaling laws for different types of process equipment (adapted from Gerber et al. (2011a)).

To confirm the validity of the methodology, a second example is shown for a compressor in Figure 2.6(b) scaled as a function of the electricity consumption in kW_e. Details for the LCIA data can be found in Steiner and Frischknecht (2007). This example provides a comparison between the conventional LCA approach for impact scaling and the proposed approach in the case where only one dataset at one particular size is known. In the figure, a second LCI dataset available for another size of the compressor is used to validate the scaling methodology. The conventional LCA approach uses a linear extrapolation ($k = 1$) to estimate the impact at any other scale. With the proposed approach based on the analogy with equipment costing, a power impact scaling law is used to extrapolate the impacts of the compressor at any size. If only one LCIA dataset is known, it is not possible to calculate the exponent directly. However, since similarity between cost and impact scaling laws is assumed, the costs exponent from the literature (Turton et al. (1998); Ulrich (1996)) can be used to estimate the impact of the compressor at any size. This gives an exponent value of 0.6.

Using the smaller compressor as reference and extrapolating the impact per kW_e – assessed by the single score of the Ecoindicator99-(h,a) method in this example – with the correlation, Figure 2.6(b) suggests that the power law for the scaling is more accurate than the linear scale-up of conventional LCA, which would assume a constant impact per kW_e. To include uncertainty parameters in this evaluation and show the relevance of the costs analogy, a Monte-Carlo simulation has been performed on the compressor at 4 and 300 kW_e using the uncertainty data included in the LCI datasets. The uncertainty linked with the linear and the power extrapolation is assumed to be constant and equal to the one of the known dataset at 4 kW_e, while the control point at 300 kW_e has its own uncertainty. The confidence interval at 95% obtained by this simulation and expressed per kW_e is shown on Figure 2.6(b), for both the dataset and the extrapolation. This highlights that the accuracy of the power law extrapolation meets the confidence interval of the datapoint, which is not the case for a conventional linear scaling.

Similar investigations have been conducted for other types of process equipment, when data were available. These include boilers and reactors. These other examples indicate as well that the use of power laws for the impact scaling following the economies of scale provides a better estimation of the impact than using a linear scaling law ($k = 1$).

All these examples indicate the validity of the assumption of the power law for the impact scaling and its analogy with the costs scaling. They however suggest that if enough data are available, it is better to directly calculate the exponents $k_{j,i}$ rather than using the costing exponent, which is actually an average approximation, since there is still a difference between the absolute value of the dataset and the costing analogy. Consequently, this option should be retained only in the case where the LCI database does not contain enough data to perform the scaling of the process equipment considered.

2.2.5   Scaling laws for process equipment

The following subsection describes the establishment of the impact scaling laws for the different types of process equipment involved in the models considered for the application case studies. The guidelines stated in Figure 2.5 are used. Table 2.1 lists these types, their associated functional parameters, and which type of correction is applied to the impact if necessary. The value of the different functional parameters and of the other parameters required to calculate the correction factors are directly taken from the thermo-economic models for the considered technologies.

Table 2.1 Process equipment types with their corresponding functional parameters and aspects taken into account for the correction factor (adapted from Gerber et al. (2011a)).

Type	Functional parameter A	Correction factor c	Scaling (Fig. 2.5)
Boiler	thermal power [kW$_{th}$]	–	Case 1
Compressor	electrical power [kW$_e$]	type (screw, axial, centrifugal)	Case 1
Engine	electrical power [kW$_e$]	used for cogeneration or not	Case 1
Filter	volume flow [Nm^3/s]	–	Case 2
Flash Drum	volume flow [Nm^3/s]	operating pressure [bar]	Case 3
Heat exchanger	exchange area [m^2]	operating pressure [bar]	Case 3
Membrane	membrane area [m^2]	–	Case 3
Pump	electrical power [kW$_e$]	operating pressure [bar]	Case 2
Reactor	volume (diameter/height) [m^3]	operating pressure [bar]	Case 3
Turbine	electrical power [kW$_e$]	–	Case 2

The established scaling laws are detailed in the paragraphs below.

Boiler

Two datasets with different thermal powers are available in ecoinvent® for an oil boiler at 10 kW$_{th}$ and at 100 kW$_{th}$ including production, assembly and disposal (Dones et al. (2007)). These datasets are then directly used to determine the exponents $k_{j,i}$ of the impact scaling law from Equation (2.12).

The mass of the boiler is calculated by Equation (2.16) for estimating the impact due to the transport. In Dones et al. (2007), masses are given for three different thermal powers of an oil boiler at 10, 100 and 1000 kW$_{th}$ and are used to calculate the exponent $k_{j,m} = 0.80$.

Compressor

Two datasets with different electrical power are available in ecoinvent® for a screw-type air compressor at 4 kW$_e$ and 300 kW$_e$ including production, assembly and disposal (Steiner and Frischknecht (2007)). These datasets are then used to determine the exponents $k_{j,i}$ of Equation (2.12).

In the case of the compressor, it is necessary to apply a correction factor c_j. Indeed, the compressor of the LCI datasets is of screw-type, while the ones considered in the present work are of centrifugal or axial type. Since the costs of an axial or centrifugal compressor are higher than for a screw-type compressor (Ulrich (1996)), a similar effect for the impact is assumed, following the analogy for emission scaling. Therefore, the correction factors corresponding to the electrical power of the two datasets are also applied to the LCI of the two datasets prior to calculating the exponents $k_{j,i}$.

The mass of the compressor for estimating the impact due to the transport is calculated by Equation (2.16). In Steiner and Frischknecht (2007), masses are given for the two different electrical powers of the screw-type compressor. The same correction factors from Ulrich (1996) were applied to the masses to estimate the exponent $k_{j,m} = 0.94$ for an axial or a centrifugal compressor.

Engine

Datasets with five different electrical powers are available in ecoinvent® for a cogeneration unit at 50 kW$_e$, 160 kW$_e$, 200 kW$_e$, 500 kW$_e$ and 1000 kW$_e$ including production, assembly and disposal (Dones et al. (2007)). For each scale, three datasets are available: one for the common components for heat and electricity production, one for the components for electricity production only and one for the components for heat production only. These datasets are then directly used to determine the exponents $k_{j,i}$ of Equation (2.12). If the engine is used for single electricity production, the components for heat are not included in the impacts. If the engine is used for cogeneration, these components are included and as well scaled with respect to the electrical power.

The specific mass of the engine is given in Dones et al. (2007) as a function of the fuel consumption. The latter is calculated by:

$$\dot{Q}^+ = \frac{\dot{E}^-}{\epsilon_{el}} \tag{2.17}$$

where $\dot{Q}^+$ is the fuel consumption of the engine, $\dot{E}^-$ its electrical power, and ϵ_{el} its electrical efficiency. Then, by replacing A_j and $A_{j,ref}$ in Equation (2.16) by $\dot{Q}^+$, the exponent is estimated to $k_{j,m} = 0.67$ for an engine.

Filter

Only one dataset is available in ecoinvent® for a central-unit filter at 600 m^3/h including production, assembly and disposal (Hässig and Primas (2007)). Though the application field of this dataset is a family house, the technology of bag filters is similar to the one used in the filter for the wood-to-SNG production unit. Therefore this dataset can be used. Since only one reference size dataset is available, it is not possible to directly calculate the exponents $k_{j,i}$ of Equation (2.10). Therefore, Equation (2.14) is used, following the costs scaling law from Ulrich (1996) for bag filters.

A similar approach to the one used for emission scaling was used to estimate the mass of the filters, since a reference mass is available in Hässig and Primas (2007).

Flash Drum

A flash drum can be assimilated to a reactor that has to be sized as a function of the gas flow rate. The diameter of the flash drum is then calculated using the following equation (Turton et al. (1998)):

$$d_j = (\frac{2}{\pi} \cdot 600 \cdot A_j)^{\frac{1}{3}} \qquad (2.18)$$

where d_j is the diameter of the flash drum and A_j is its gas volumetric flow rate. Then, the same method than for the reactor detailed below is applied to calculate the mass of stainless steel and the same correction factor is applied for pressure.

Heat exchanger

Since detailed design data for a shell and tube heat exchanger are available at different sizes and that no datasets are available at different sizes in ecoinvent®, the design data are used to directly calculate the quantities of materials. The heat exchanger is composed by two different materials: stainless steel for the tubes and the shell, and unalloyed steel for the other parts of the heat exchanger. Equivalences are available for the production of these two materials in ecoinvent®. For each material, a mathematical expression is developed to scale its quantity as a function of the exchange area, based on the available design data. The amount of stainless steel was found to follow a linear scaling law of the type:

$$\frac{m_j}{m_{j,ref}} = \frac{A_j}{A_{j,ref}} \qquad (2.19)$$

where m_j is the mass of stainless steel to be calculated $m_{j,ref}$ the one of a reference heat exchanger and A_j is the exchange area. The amount of unalloyed steel was found to follow a similar law to Equation (2.16). The exponent $k_{j,m}$ was estimated to 0.59. These masses are then used to calculate the LCI of the associated equivalent materials, with Equation (2.15).

For the cost estimation of a heat exchanger, it is also necessary to apply a correction factor for the pressure (Turton et al. (1998)). Following the analogy between cost and impact estimation, the correction factor c_j for the pressure is also applied to the impact scaling, and the correlations from Turton et al. (1998) are used to correct the emissions of the LCI.

The masses are also used to calculate the impacts due to the transport and disposal. For the disposal, it is assumed that 98% of the steel can be recycled. This is the assumption made by Felder and Dones (2007) for the recycling of metal catalysts. The remaining 2% are assumed to be disposed in sanitary landfill as inert material. The impacts from the assembly and the disassembly are neglected.

Membrane

There is no specific dataset available in ecoinvent® for a membrane. However, since this one is made from polymer, an equivalence for polymer in a sheet form is used. Then, the mass of polymer is calculated by the following equation:

$$m_j = \rho_j \cdot A_j \cdot \theta_j \qquad (2.20)$$

where m_j is the mass of the membrane, ρ_j is the density of the polymer, A_j is the exchange area of the membrane, and θ_j is the membrane thickness.

The impacts due to the transport and disposal are also calculated using the quantity of polymer. For the disposal, the membranes are assumed to be incinerated.

Pump

Only one dataset is available for a pump at $0.04\ kW_e$ in ecoinvent® including production, assembly and disposal (Dones et al. (2007)). It is therefore not possible to directly calculate the exponents $k_{j,i}$ with Equation (2.12). Therefore, Equation (2.14) was used, following the cost scaling correlation from Turton et al. (1998) for centrifugal pumps.

For the cost estimation of a pump, it is also necessary to apply a correction factor for the pressure (Turton et al. (1998)). Following the analogy between cost and impact estimation, the correction factor c_j for the pressure is also applied to the impact scaling, and the correlations from Turton et al. (1998) are used to correct the emissions of the LCI before LCIA.

For the mass, a similar approach to the one used for the emission scaling is used, since a reference mass is available from Dones et al. (2007).

Reactor

Since no dataset is available in ecoinvent® for a reactor, it is necessary to use design data. The reactor is therefore assimilated to a cylinder made of stainless

steel. An equivalence for stainless steel is available in the LCI database, and
the amount of stainless steel is calculated by:

$$m_j = \rho_j \cdot \theta_j \cdot \pi \cdot (d_j \cdot h_j + 2 \cdot (\frac{d_j}{2})^2) \qquad (2.21)$$

where m_j is the mass of the reactor, assumed to be fully made of stainless steel,
ρ_j is the density of stainless steel, θ_j is the thickness of the reactor wall, d_j is
the diameter, and h_j is the height of the reactor.

For the cost estimation of a reactor, it is also necessary to apply a correction
factor for the pressure (Turton et al. (1998)). Following the analogy between
costs and impacts estimation, the correction factor c for the pressure is also
applied to the impact scaling, and the correlations from Turton et al. (1998)
are used to correct the emissions of the LCI.

The mass of stainless steel is also used to calculate the impacts due to transport
and disposal. For the disposal, it is assumed that 98% of the steel can be
recycled, and therefore are leaving the system. This is the assumption made by
Felder and Dones (2007) for the recycling of the metal catalysts. The remaining
2% are assumed to be disposed in the sanitary landfill as inert material. The
impacts from the assembly and the disassembly are neglected.

Turbine

Only one dataset is available for a gas turbine at 10 000 kW$_e$ in ecoinvent®,
including production, assembly and disposal (Dones et al. (2007)). Therefore,
it is not possible to directly calculate the exponents $k_{j,i}$ with Equation (2.12).
Equation (2.14) was used instead, following the cost scaling correlation from
Ulrich (1996) for radial turbines.

For the mass, a similar approach to the one used for the emission scaling is
used, since a reference mass is available from Dones et al. (2007).

2.2.6 Multi-period formulation

The computational framework described in Figure 2.1 is valid for solving problems where the quantities of available resources or of energy services supplied
are constant in time.

However, variations in the resource availability or in the energy services requirements may occur in time. District heating is a good example, since the
requirements of a given residential area or of a city present a seasonal variation
during the year. This one can be divided in independent periods with fixed
operating conditions and requirements (Girardin et al. (2010)). It is therefore
necessary to adapt the computational framework of Figure 2.1 to solve such
problems.

Without considering the possibilities of heat or mass storage, the problem to be solved is divided in n_p independent periods. Thus the MILP slave sub-problem, representing the system operating conditions for a given period, is solved for each one of the periods, as illustrated in Figure 2.7. The economic, thermodynamic and environmental indicators are calculated for each period, and reflect thus the variations in process configuration and efficiency.

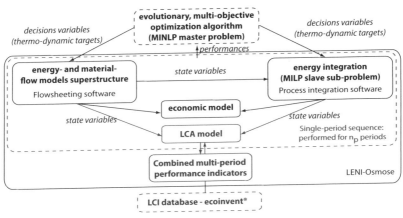

Figure 2.7 Computational framework for system simulation and design accounting for multi-period aspects.

Specifically concerning the LCA in multi-period and the calculation of the LCI, a distinction has to be made between the elements of the LCI belonging to the operation phase and the ones belonging to the construction or end-of-life phase. This is done by analogy with the economic indicators, for which a distinction is made between the operating costs or revenues and the investment linked with the equipments. While the costs and revenues are added to each other for each period, the equipments are sized to operate in the conditions representative of any of the considered periods. Hence, the period where an equipment operates at its maximum capacity determines its size and thus its investment.

These interactions between the operation phase, and the construction and end-of-life stages, are illustrated by Figure 2.8.

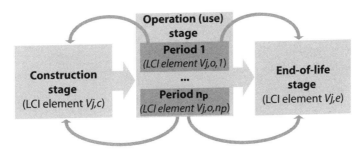

Figure 2.8 Interactions between the different stages of the life cycle when multi-period aspects are considered.

For instance, a LCI element $V_{j,c}$ associated with the construction stage – e.g. a turbine to produce electricity – or $V_{j,e}$ associated with the end-of-life stage – e.g. the percentage of disposed material of the turbine – might have their size decided by another LCI element $\dot{V}_{j,o,p}$ associated with the operation phase – e.g. the electricity produced by the turbine at each period. The sizes of $V_{j,c}$ and $V_{j,e}$ are then fixed by the maximal value of the element $\dot{V}_{j,o,p}$ over all the periods:

$$V_{j,c}, V_{j,e} \sim max(\dot{V}_{j,o,p})_{n_p} \tag{2.22}$$

Therefore, accounting for the life cycle perspective and for the multi-period aspects of the system operation, the final impacts per functional unit for each impact category of the chosen impact assessment methods are given by:

$$I_l^{FU} = \frac{\sum_{p=1}^{n_p} \sum_{j=1}^{n_{j,o}} \dot{IO}_{j,p,l}(x_d) \cdot t_p \cdot t_{yr} \cdot r_o}{FU_{tot}}$$

$$+ \frac{\sum_{j=1}^{n_{j,c}} IC_{j,p,l}(\dot{IO}_l, x_d)_{n_p} + \sum_{j=1}^{n_{j,e}} IE_{j,p,l}(\dot{IO}_l, x_d)_{n_p}}{FU_{tot}} \tag{2.23}$$

where $\dot{IO}_{j,p,l}$ is the impact associated with impact category l due to the operation phase for period p of the LCI element j, $n_{j,o}$ being the number of LCI elements associated with operation phase, $IC_{j,p,l}$ is the impact due to the construction phase of the LCI element j, $n_{j,c}$ being the number of LCI elements associated with construction phase, $IE_{j,p,l}$ is the impact due to the end-of-life phase of the LCI element j, $n_{j,e}$ being the number of LCI elements associated with end-of-life phase, t_p is the time associated with period p, t_{yr} the lifetime of the system, r_o the yearly percentage of operation of the system and FU_{tot}, the total functional unit quantity. If FU_{tot} is a quantity varying over all the different periods – e.g. the district heating requirements of a residential area – Equation (2.7) for constant operating conditions becomes:

$$FU_{tot} = \sum_{p=1}^{n_p} \dot{FU}_p(x_d) \cdot t_p \cdot t_{yr} \cdot r_o \tag{2.24}$$

where $\dot{FU}_p$ is the quantity of functional unit involved during period p.

Examples of multi-period application case studies are provided in Chapters 4 and 5.

2.3   Industrial ecology and supply chain synthesis

As explained in the previous sections, the methodology for integrating the LCA
in the system design requires to extend the decision perimeter of the process
flowsheet for including the auxiliary materials, the emissions, the logistics and
the generated waste. Accounting for off-site emissions involves the use of a
LCI database. However, such databases assume a fixed process or supply chain
for each product or waste in which pre-defined fractions of each process represent the market share of each option. Therefore, process design options may
be biased due to the choices of the options in the process chains. Moreover,
the systematic identification of industrial ecology possibilities, dealing with the
closing of material loops by waste and co-product recycling, is not possible
with such databases and with the previously presented methodology, considering only energy and not mass integration. These issues are particularly critical
for the application of the methodology to complex energy systems, such as
urban systems, where multiple resources or waste can be used with multiple
technologies to produce multiple energy services. In order to identify the maximal potential for impact mitigation, it is necessary to account for the possible
recyclings and to re-synthesize the supply chains down to a certain level where
the average market conditions can not be influenced anymore.

The systematic identification of industrial ecology possibilities and the synthesis
of supply chains using the LCI database ecoinvent® is possible by extending
the computational framework and the methodology presented in Figure 2.1 and
2.7. The modified computational framework is presented in Figure 2.9.

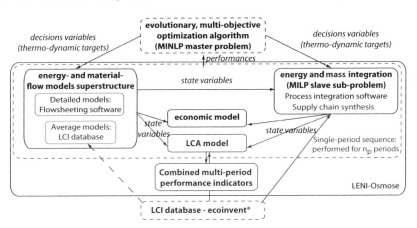

Figure 2.9 Computational framework for system simulation and design accounting for multi-period aspects and supply chain synthesis (adapted from Gerber et al.
(2012)).

Using combined mass and energy integration, the process integration step is
extended to the overall system integration, including the supply chains of auxiliary materials, logistics and resources that are locally necessary to the process.

The objective of the MILP slave sub-problem is the minimization of the operating cost, or the minimization of the impact, or the minimization of the operating costs including environmental taxes. For these last two objectives, it is necessary to calculate the environmental impact of the different processes embedded in the superstructure prior the post-calculation phase, and a link is therefore created between the LCI database and the process integration software to systematically extract these impacts. Moreover, the decision system extension involves to include models in the superstructure of average technologies or resources for which detailed models with a flowsheeting software are not necessary. In this case, the data of average technologies in the LCI database are sufficient in precision and can be used as a basis for writing simple models.

2.3.1   Decision system extension

The general concept developed to generate a superstructure that allows to account for supply chain synthesis in the process integration is described in Figure 2.10, which is a more detailed description of the models embedded in the superstructure of Figure 2.9. A distinction is made between the overall system, representing the real system to be considered for calculating the environmental impacts or the total costs, and the action system. This extended action system represents the overall space where decisions can be effectively taken in order to decrease the impacts or the costs. It contains thus a superstructure of models that are not only limited to process flowsheets for the conceptual design of technologies, but it also embeds average technologies and resources that are involved in the supply chain and for which decisions are possible.

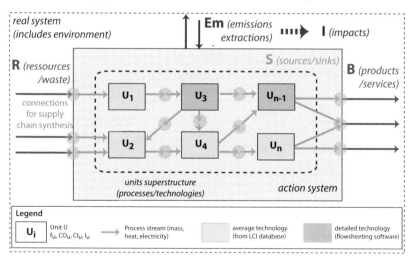

Figure 2.10 General concept for supply chain synthesis including data from LCI database (adapted from Gerber et al. (2013)).

The action system contains a superstructure of units representing processes or technologies (U), used to synthesize the supply chain necessary to obtain a given amount of final products (B), or to convert waste or available resources

(R). A unit is constituted of a series of sources and sinks (S), representing material and energy streams that are necessary inputs or product outputs. The units are either detailed models using flowsheeting software, or simple models based on the average technologies of the LCI database. Each unit generates emissions and extractions (Em) to and from the environment, creating impacts (I). Each unit has an associated utilization factor (f_u), to be determined by solving the MILP slave sub-problem at the process integration step, an associated operating cost (CO_u), an investment cost (CI_u) and an associated impact (I_u). The supply chain, which aims at determining the optimal values of the f_u, is then synthesized by establishing matches between sources and sinks, with respect to the goal of the study and to the definition of the functional unit (FU), according to the LCA methodology. This FU can be a given quantity of a product or a service to be produced, or available resources to be converted or waste to be treated.

This general concept is used as a basis to disaggregate in a first step the LCI database ecoinvent® in unit processes that represent models of average technologies, with associated material and energy flows, and emissions and extractions. In a second step, the models are used to re-synthesize the supply chains accounting for the mass balance constraints, identifying automatically the possible recyclings, and minimizing the impacts on the environment or the costs.

2.3.2 LCI database disaggregation

To disaggregate the LCI database ecoinvent® in models of average resources or technologies, a distinction has to be made between the unit processes termed here as 's-type', which are fully included in the superstructure of the action system, having both sources and sinks, and the unit processes of 'r-type' and 'p-type', which are at the interface of the action system and the environment. r-type unit processes are used for a resource delivery, and p-type unit processes for a product sink. The r-type and p-type units represent the limits of the system under which decisions have to be taken. These three types are illustrated in Figure 2.11.

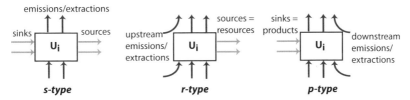

Figure 2.11 Differences between the s-type, r-type and p-type unit process (adapted from Gerber et al. (2011b)).

The level of cut-off at which a unit process becomes a r- or p-type has to be defined considering the level at which no control is possible anymore on the processes, representing the average technology or resources available on the market. It is relatively easy to perform this operation with the data from ecoinvent®, since the database offers two different formats for the information linked with the unit processes:

1) UPR format (unit process), used for s-type processes, which contains the inputs and outputs in terms of material and energy flows, representing the sources and sinks. It contains as well the emissions and extractions generated by the process only, without upstream and downstream emissions.

2) Aggregated LCI format, used for r- and p-type unit processes, containing the cumulative emissions and extractions of the upstream or downstream processes.

Once the disaggregation is performed, the different processes constitute the superstructure of the average technologies and resources that are used in the next step to re-synthesize the optimal supply chain.

2.3.3 MILP formulation for supply chain synthesis

To synthesize the optimal supply chain and account for the systematic recyclings, a MILP model is used for each independent operation period. The constraints are put on the mass balance for each process or technology, to ensure that each source for a process finds another process as a sink, except where a certain quantity of a product is wanted or of a resource has to be converted:

$$\forall s = 1...n_s : \sum_{u=1}^{n_u} f_{u,p} \cdot \dot{s}^+_{s,u,p}(x_d) = 0 \qquad (2.25)$$

$$\forall r = 1...n_r : \sum_{u=1}^{n_u} f_{u,p} \cdot \dot{r}^+_{r,u,p}(x_d) = \dot{R}_r \qquad (2.26)$$

$$\forall b = 1...n_b : \sum_{u=1}^{n_u} f_{u,p} \cdot \dot{b}^+_{b,u,p}(x_d) = \dot{B}_b \qquad (2.27)$$

where $f_{u,p}$ is the utilization fraction at period p of the process or technology u of the superstructure for the considered action system, n_u being the number of processes embedded in the superstructure, $\dot{s}^+_{s,u,p}$ is the quantity of the source s consumed by the process u, a negative sign meaning a production. $\dot{s}^+_{s,u,p}$ is function of the decision variables x_d of the master MINLP problem of Figure 2.9. $\dot{r}^+_{r,u,p}$ is the quantity of resource r consumed by process u, and $\dot{b}^+_{b,u,p}$ is the quantity of product b. $\dot{R}_r$ and $\dot{B}_b$ are the required quantities of resource r or product b that have to be converted or produced, and are defined according to the problem to be solved. Using this layer approach for each source/sink, product and resource allows for systematically identifying the potential recyclings within the action system. The $f_{u,p}$ are the decision variables of the slave

MILP sub-problem. The emissions and extractions calculation for the LCI is as well included in the MILP formulation:

$$\forall i = 1...n_i : \sum_{u=1}^{n_u} f_{u,p} \cdot \dot{em}_{i,u,p}^{+}(x_d) = \dot{Em}_{i,p} \qquad (2.28)$$

where $\dot{em}_{i,u,p}^{+}$ is the elementary flow i of the LCI emitted or extracted by the process u. Thus, a linear approximation of the impact of the system for its operation is calculated during the process integration step by using Equations (2.3) and (2.4) for the impact assessment.

This impact can be used to solve the MILP slave sub-problem, in order to integrate environmental taxes in the system operating costs or to minimize the impact. The definition of operating costs becomes thus:

$$Min \ \dot{CO}_p = (\sum_{u=1}^{n_u} f_{u,p} \cdot (\dot{CO}_{u,p} + \dot{I}_{u,p} \cdot c_I)$$
$$+ \sum_{r=1}^{n_r} (\dot{R}_{r,p} \cdot c_r + \dot{I}_{r,p} \cdot c_I) + \dot{E}_p^{+} \cdot c_{e+} - \dot{E}_p^{-} \cdot c_{e-})(x_d) \qquad (2.29)$$

where $f_{u,p}$ is the utilization factor of unit u at period p, $\dot{CO}_{u,p}$ is its operating cost, $\dot{I}_{u,p}$ its impact, c_I is the specific environmental tax associated with the impact, $\dot{R}_{r,p}$ is the consumption of resource r at period p, c_r is its buying price, $\dot{I}_{r,p}$ is its impact, $\dot{E}_p^{+}$ and $\dot{E}_p^{-}$ are the consumed and produced electricity by the system, respectively, and c_{e+} and c_{e-} are the specific costs for electricity buying and selling from and to the grid, respectively.

If the impact is used as the objective of the MILP slave sub-problem, the formulation becomes:

$$Min \ \dot{I}_p = (\sum_{u=1}^{n_u} f_{u,p} \cdot \dot{I}_{u,p} + \sum_{r=1}^{n_r} \dot{I}_{r,p} + \dot{I}_{E_p^{+}} - \dot{I}_{E_p^{-}})(x_d) \qquad (2.30)$$

where $\dot{I}_{u,p}$ is the impact of unit u at period p, $\dot{I}_{r,p}$ the impact of resource r, and $\dot{I}_{E_p^{+}}$ and $\dot{I}_{E_p^{-}}$ are the impact of electricity import and export from and to the grid, respectively.

For each independent period, both equations (2.29) and (2.30) are as well submitted to the constraints of the heat cascade (Linnhoff et al. (1982); Maréchal and Kalitventzeff (1998)), given by Equations (2.31) and (2.32) for each temperature interval k and to the constraints for electricity consumption and exportation, given by Equations (2.33) to (2.35):

$$\sum_{u=1}^{n_u} f_{u,p} \cdot \left(\sum_{h_k=1}^{n_{u_{h,k}}} \dot{Q}_{u_{h,k,p}} - \sum_{c_k=1}^{n_{u_{c,k}}} \dot{Q}_{u_{c,k,p}} \right) + \dot{R}_{k+1,p} - \dot{R}_{k,p} = 0 \quad \forall k = 1..., n_k \quad (2.31)$$

$$\dot{R}_{1,p} = 0 \quad \dot{R}_{nk+1,p} = 0 \qquad \dot{R}_{k,p} \geq 0 \quad \forall k = 2..., n_k \qquad (2.32)$$

$$\sum_{u=1}^{n_u} f_{u,p} \cdot \dot{E}_{u,p}^{+} + \dot{E}_p^{+} - \sum_{u=1}^{n_u} f_{u,p} \cdot \dot{E}_{u,p}^{-} \geq 0 \qquad (2.33)$$

$$\sum_{u=1}^{n_u} f_{u,p} \cdot \dot{E}_{u,p}^{+} + \dot{E}_p^{+} - \dot{E}_p^{-} - \sum_{u=1}^{n_u} f_{u,p} \cdot \dot{E}_u^{-} = 0 \qquad (2.34)$$

$$\dot{E}_p^{+} \geq 0 \qquad \dot{E}_p^{-} \geq 0 \qquad (2.35)$$

where $\dot{Q}_{u_{h/c,k},p}$ is the thermal heat load of the hot stream h or the cold stream c in the unit u in the temperature interval k, in kW_{th} during period p. The hot and cold streams are calculated for the nominal size of each resource, technology or service included in the superstructure. An inlet and outlet temperature and a thermal load are associated to each stream. The temperatures of all streams define n_k intervals. $\dot{R}_{k,p}$ is then the cascaded heat from the temperature interval k to the lower ones, in kW_{th}, by starting with the higher interval. $\dot{E}_{u,p}^{+}$ and $\dot{E}_{u,p}^{-}$ are the consumed and exported electricity in kW_e by unit u, respectively. The constraints on $\dot{E}_p^{+}$, $\dot{E}_p^{-}$ and $\dot{R}_{k,p}$ guarantee the thermodynamic feasibility.

The utilization factors f_u are limited by a minimum and a maximum value. The associated integer variables y_u define if the unit u is added to the system ($y_u = 1$) or not ($y_u = 0$):

$$y_u \cdot f_{u,min} \leq f_u \leq y_u \cdot f_{u,max} \qquad (2.36)$$

In the above equations, the decision variables of the MILP slave sub-problem are written in regular typesetting.

Depending on the problem to be solved, a modified formulation of the heat cascade including heat exchange restrictions can as well be used (Becker and Maréchal (2012a)).

An example of application of the methodology for the synthesis of an urban energy conversion system is provided in Chapter 5.

CHAPTER 3

Application to thermochemical wood conversion

3.1 Combined SNG, heat and electricity production from lignocellulosic biomass

Among the candidate renewable resources that could play in important role for the energy future of Switzerland, residual woody biomass from forest exploitation has an under-exploited potential nowadays, mostly for heating purposes (OFEN (2010a)).

However, the transport sector contributes to about one third of the CO_2 emissions in Switzerland, and suffers currently from a lack of alternatives to fossil fuels (OFEV (2011)). There is thus a growing interest for the production of second-generation biofuels based on indigenous lignocellulosic biomass, which appears to be a more sustainable alternative than first-generation biofuels based on food crops, as suggested by Zah et al. (2007).

One of the candidate technologies for the production of fuel from lignocellulosic biomass is the thermochemical conversion of biomass to Synthetic Natural Gas (SNG). SNG can be used then to substitute conventional fossil natural gas for the production of multiple energy services, such as transport, electricity, heating, or cogeneration of electricity and heat. This biomass conversion process requires water and produces biogenic methane and carbon dioxide, which has to be removed from the final product. Taking the typical wood chemical composition, this conversion can be represented by the following chemical reaction (Gassner and Maréchal (2009b)):

$$CH_{1.35}O_{0.63} + 0.3475\ H_2O \longrightarrow 0.51125\ CH_4 + 0.48875\ CO_2,$$
$$\Delta \tilde{h}_r^0 = -10.5\ \mathrm{kJ\ mol}_w^{-1} \tag{3.1}$$

where $\Delta\tilde{h}_r^0$ is the standard heat of reaction, per mole of wood. The overall equation being exothermal, the additional available heat can be used in a steam cycle to produce electricity or for heating purposes.

At the present time, the thermochemical conversion of lignocellulosic biomass in multiple energy services is still an emerging technology, characterized with a large number of technological options and potential operating conditions (Stucki et al. (2010); Gassner (2010)). The technology is for the moment still at the stage of pilot-scale process and the final commercial design is therefore not fixed yet. This makes it an interesting example to study the effects of the potential variations in the process design and configuration on the environmental impacts by the developed methodology presented in this work.

Moreover, a LCA following the conventional methodology has already been conducted on such a pilot-scale process (Felder and Dones (2007)), making thus possible the comparison with the developed methodology. This is why the thermochemical conversion of lignocellulosic biomass is presented here as a first case study to illustrate the interest of integrating the LCA in a process design and integration framework.

3.2   Models description

Thermo-economic models for the different technologies and process configurations for the thermochemical conversion of lignocellulosic biomass to SNG have been developed by Gassner (Gassner and Maréchal (2009a,b); Gassner (2010)), who performed as well the systematic identification of the optimal process configurations considering thermo-economic criteria. These models have been used as a basis for the present study and have been extended to LCA models according to the methodology presented in section 2.2.

3.2.1   Thermo-economic model

Figure 3.1 presents the process superstructure from Gassner and Maréchal (2009b) for the thermochemical conversion of lignocellulosic biomass to SNG.

Five or six processes are necessary to convert lignocellulosic biomass into SNG of a sufficient quality to be injected in the gas grid. At each step, different potential technologies can be used (Gassner (2010)). These stages and their corresponding candidate technologies are briefly described below.

Drying. This first step is necessary to remove part of the high moisture content of the woody biomass, which otherwise decreases the performance of the gasification process. Potential technologies that can be used are air or steam drying.

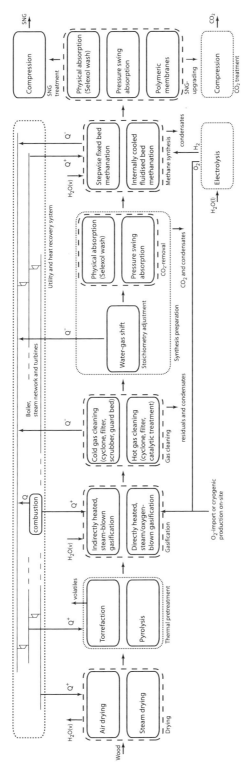

Figure 3.1 Process superstructure for the conversion of lignocellulosic biomass to SNG. Dashed lines assemble investigated alternatives for different process sections and dotted lines indicate optional units (reproduced from Gassner and Maréchal (2009b) with permission).

Pretreatment. A thermal pretreatment can be done in addition to drying prior the gasification process, but is optional. This can be performed either by torrefaction or pyrolysis.

Gasification. The gasification is an endothermal process, operating around 800-850°C and breaking down the larger molecules of the biomass in a producer gas, which is a mixture of hydrocarbons, hydrogen, carbon monoxide and carbon dioxide. Different technologies – directly heated circulating fluidized bed (CFB), indirectly heated fast internally circulating fluidized bed (FICFB), pressurized indirectly heated fast internally circulating fluidized bed (pFICFB) – and gasifying agents – air, steam or oxygen – can be used. In indirectly heated gasification, the heat is provided to the gasifier by a separate combustion chamber using part of the biomass, while in direct gasification the heat is directly provided within the gasifier by oxidizing part of the biomass with pure oxygen. These two processes can be potentially operated under pressure or not.

Gas cleaning. Since the producer gas contains several impurities, such as tars, ashes, sulphur and others, it has to be cleaned of its impurities prior the methanation stage, to avoid catalyst poisoning. Conventional technology is cold gas cleaning (CGC), with a filter, a scrubber and guard beds. Hot gas cleaning (HGC) is another alternative, which does not require a scrubbing step.

Methane synthesis. The aim of this step is to increase the methane fraction in the producer gas, which contains high fractions of carbon dioxide and hydrogen. The process operates around 300-400°C under pressure or not, and is a highly exothermic process. At the end of this stage, the produced gas is mostly a mixture of methane and carbon dioxide.

Purification. This step aims at upgrading the SNG by removing the carbon dioxide from the gas to meet the standards for injection in the gas grid. The three most suitable technological options are the physical absorption, the pressure swing absorption (PSA) and membrane processes.

A more detailed description of the process modeling can be found in Gassner and Maréchal (2009b) and Gassner (2010).

The thermochemical conversion of lignocellulosic biomass allows for the combined production of SNG, electricity and heat that can be used as process heat or for district heating. For the present application case study, only the combined SNG and electricity production are considered, since the issue of district heating and of its seasonal variations is addressed in the next chapters. Thus, the computational framework of Figure 2.1, with constant operating conditions for the process, is used.

3.2.2   Life Cycle Assessment model

The LCA model for the different components involved in the thermochemical conversion of lignocellulosic biomass in energy services is established by following the methodology described in Figure 2.2.

Functional Unit definition

The question to be answered with this application case study relates to the quantification of the environmental impacts associated with the production of SNG from lignocellulosic biomass and how they are related to the changes in process design and configuration. Since the interest of the case study is to focus on the influence of the conversion process design, the different possible allocations of produced SNG that would be potentially in transportation, heating or electricity production, are not considered. Therefore, it would seem at first natural to adopt a cradle-to-gate approach and consequently to choose 1 MJ of produced SNG as the functional unit (FU). However, this FU is not suitable for the following reasons.

Due to electricity cogeneration from excess heat, both SNG and electricity are potential products whose relative yields depend on the process design. In an environomic optimization, adopting a strict cradle-to-gate approach with 1 MJ of SNG as FU leads to paradoxal results: since only the benefit of electricity generation from renewables is considered by an asymetric substitution of the multiple products, minimizing the impact of the SNG production is obtained by maximizing the electricity production to the expense of SNG, whose beneficial effect is not accounted for. Technical inefficiency and waste of renewable resources would thus be promoted. This problem is directly linked to the strict single-product based cradle-to-gate approach using 1 MJ of SNG as the FU.

The second reason is that the SNG production from lignocellulosic biomass is divided in two phases: the wood chips production and its conversion to SNG. The latter is thereby of particular interest, since it is mainly at this step that engineering decisions that impact the environment are taken. Using 1 MJ of wood as FU instead of 1 MJ of SNG fixes the impact of wood production to a constant value that does not change depending on the considered scenario, while keeping the impact of logistics, which depends on the average collection distance related to the plant size. Therefore, the effect of the process design and of the conversion efficiency is highlighted, and the impact of the SNG produced is considered by substituting its production as it is the case for the electricity.

For these reasons, the chosen FU is 1 MJ of wood entering the conversion process, considering that the function of the SNG plant is to convert wood into useful energy services – i.e. SNG and electricity. Therefore, the substitution for all the produced energy services is included in the system boundaries. Regarding the SNG production, no special assumption regarding the allocation of the produced SNG is thereby required.

Life Cycle Inventory model

The three different element types involved in the LCI – flows of the process flowsheet, flows of auxiliary materials and process equipment – are identified and quantified, following the methodology described in section 2.2.3. In order to account for the off-site emissions, the LCI database ecoinvent® is used to find equivalences for each LCI element.

Figure 3.2 displays the flows of the process flowsheet and of the auxiliary materials that are included in the system boundaries. They have been identified on the basis of the inventories of Felder (2004) and Felder and Dones (2007). The flows of auxiliary materials and emissions are included, but also the flows of the process flowsheet. The step of the SNG production at which they occur is as well displayed on the figure, which is a simplified schematic representation of the process detailed by Gassner and Maréchal (2009a).

The figure shows that the use of LCA requires to extend the process flowsheet to include in the LCI the auxiliary materials that are not accounted for in the thermo-economic calculations, but that are nevertheless of environmental significance.

The substitution of the produced energy services – i.e. SNG and electricity – is included in the LCA system boundaries, in order to account for the process efficiency. For SNG, this includes the avoided fossil natural gas extraction and the avoided fossil CO_2 emissions from the combustion of fossil natural gas. For electricity, Felder and Dones (2007) assume that consumed electricity comes from the actual Swiss mix, including the imports. The same hypothesis is therefore made for substituting the produced electricity. At the time of the study, which was before the Fukushima nuclear accident and the decision of the Swiss Federal Council to abandon progressively nuclear power, this assumption was realistic, since the impacts of the actual Swiss mix including the imports are dominated by the impacts of nuclear power and of electricity import produced from conventional fossil resources. The equivalences in ecoinvent® used for the LCI flows are available in Annex A.1.

The inventory of the process equipment is based on the equipment that is rated and quoted in the economic model, for consistency between the thermo-economic and the LCA models. Following the methodology described in section 2.2.3, a specific type is assigned to each equipment.

According to the methodology, for the flows already included in the thermo-economic model, the quantities are directly extracted from the model results. For the process equipment, the methodology presented in section 2.2.4 is used for impact scaling. For the auxiliary materials, the formulation is developed on a case-by-case basis following Equation (2.8), since for the case of SNG production, the auxiliary flows are related to the operation phase. The construction and the end-of-life concern here the process equipment of the plant,

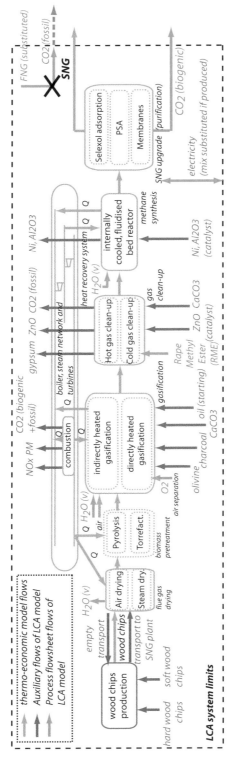

Figure 3.2 LCI elements for the conversion of lignocellulosic biomass to SNG and electricity (adapted from Gerber et al. (2011a)).

and the impacts related to these life cycle stages are specifically addressed by the methodology of subsection 2.2.3. Table 3.2, 3.3 and 3.4 summarize all the flows and process equipment included in the LCI for the life cycle of one plant for biomass transport and conversion, with the associated parameters that allow for calculating them as a function of the process configuration. Their reference quantities v_j are as well given when necessary. Table 3.1 provides the description of the parameters used in the equations to calculate the LCI elements.

Table 3.1 Description of the parameters used to express the LCI elements.

Abbreviation	Description	Unit	Range
$\dot{Q}_w^+$	Process scale, expressed as thermal capacity of input wood	kW_{th}	5000-200 000
$\dot{m}_w^+$	Mass flow rate of wood at the process inlet	kg/s	from model
x_d	All decision variables of the MINLP master MOO problem	–	–
ω	Wood humidity after drying	–	0.1-0.3
T_{gas}	Gasification temperature	°K	1073-1173
P_{gas}	Gasification pressure	bar	1.15-20.15
x_{CH_4}	CH_4 recovery in CO_2 removal	–	0.95-0.99
$x_{d,memb}$	Parameters specific to the membranes model	–	–
T_{dr}	Drying temperature	°K	453-513
T_{pyr}	Pyrolysis temperature	°K	723-1023
T_{tor}	Torrefaction temperature	°K	623
$T_{meth,in}$	Methanation inlet temperature	°K	573-673
$T_{meth,out}$	Methanation outlet temperature	°K	573-673
P_{meth}	Methanation pressure	bar	1.15-30.15

Wood production. The mass flow rate of dried wood chips required by the process is calculated by the thermo-economic model as a function of the thermal capacity of input wood of the process $\dot{Q}_w^+$, which determines the process scale, and of the Lower Heating Value (LHV) of the biomass, calculated as a function of the biomass chemical composition, which is an input parameter of the model. Since the wood chips from residual wood from forest industry can be either from hard wood – i.e. from Angiosperms tree species – or soft wood – i.e. from Gymnosperms tree species – a base mix of 50% hard wood and 50% soft wood is taken, since both kinds of wood are currently commercialized in Switzerland (Werner et al. (2003); Spielmann et al. (2007)). The mass flow rate of wood chips has to be converted in volume for consistence with the units used in the ecoinvent® equivalences:

$$\dot{V}_{w,j}^+ = \beta_j \cdot \frac{\dot{m}_w^+}{\rho_{w,j,dry}} \tag{3.2}$$

Table 3.2 Summary of the material and energy flows included in the LCA model for biomass production and logistics, and of the parameters used for their scaling and adaptation to the configuration.

Name of LCI element & Unit	Process stage	Techno-logy	Func-tional param-eters	Reference quantity v_j	Source
Hard wood chips [m^3]	Wood chips production	all	$\dot{Q}_w^+$	0.5 $\dot{m}_w^+$	Gassner and Maréchal (2009b); Gerber et al. (2011a)
Soft wood chips [m^3]	Wood chips production	all	$\dot{Q}_w^+$	0.5 $\dot{m}_w^+$	Gassner and Maréchal (2009b); Gerber et al. (2011a)
Transport from forest to SNG plant [tkm]	Wood chips transport	all	$\dot{Q}_w^+$	see Equation (3.3)	Stucki et al. (2010)
Empty transport from SNG plant to forest [km]	Wood chips transport	all	$\dot{Q}_w^+$	see Equation (3.3)	Stucki et al. (2010)

where $\dot{V}_{w,j}^+$ is the volume flow rate of wood fraction j, β_j is the fraction of wood fraction j in the wood chips mix, and $\rho_{w,j,dry}$ is the dry wood density at 0% humidity of the fraction j. For hard wood, data from Werner et al. (2003) give values of $\beta_j = 0.5$ and $\rho_{w,j,dry} = 239$ kg/m^3, and for soft wood values of $\beta_j = 0.5$ and $\rho_{w,j,dry} = 169$ kg/m^3.

Wood logistics. For the biomass logistics, which includes the transport from forest to SNG plant and the empty transport back to the forest, data based on Geographic Information Systems (GIS) for a specific location, Eclépens, in Switzerland, are used. This allows for evaluating the average shortest transport distance from the forest to the SNG plant for a given process scale $\dot{Q}_w^+$ (Stucki et al. (2010)). The average transport distance between forest and SNG plant is thus calculated by:

$$D = d_1 \cdot \dot{Q}_w^{+\ d_2} \tag{3.3}$$

where d_1 and d_2 are constants estimated to 0.0535 km/kW$_{th}$ and 0.58, respectively, using the GIS data.

Water. Since water is required and produced at different stages of the process, as shown in Figure 3.2, its total required quantity is based on the water balance, which is already performed by the thermo-economic model. This allows to account for potential recycling, minimizing thereby external water consumption.

Table 3.3 Summary of the material and energy flows included in the LCA model for SNG conversion process, and of the parameters used for their scaling and adaptation to the configuration.

Name of LCI element & Unit	Process stage	Tech-nology	Functional parameters	Reference quantity v_j	Source
Water for process [kg]	Balance	all	$\dot{Q}_w^+, x_d$	from model	Gassner and Maréchal (2009b)
Olivine [kg]	Gasif.	all	$\dot{Q}_w^+, T_{gas}, P_{gas}$	$4.62 \cdot 10^{-4} \mathrm{kg/m^3/s}$	Felder and Dones (2007)
Charcoal [kg]	Gasif.	all	$\dot{Q}_w^+, T_{gas}, P_{gas}$	$2.6 \cdot 10^{-4} \mathrm{kg/m^3/s}$	Felder and Dones (2007)
Calcium carbonate [kg]	Gasif.	all	$\dot{Q}_w^+, T_{gas}, P_{gas}$	$2.6 \cdot 10^{-4} \mathrm{kg/m^3/s}$	Felder and Dones (2007)
Starting oil [kg]	Gasif.	all	$\dot{Q}_w^+, T_{gas}, P_{gas}$	$5.67 \cdot 10^{-6} \mathrm{kg/m^3/s}$	Felder and Dones (2007)
Oxygen (cryogenic) [kg]	Gasif.	CFB	$\dot{Q}_w^+, T_{gas}, P_{gas}$	from model	Gassner and Maréchal (2009b)
Solid waste disposal [kg]	Gasif.	all	$\dot{Q}_w^+, T_{gas}, P_{gas}$	$9.88 \cdot 10^{-4} \mathrm{kg/m^3/s}$	Felder and Dones (2007)
NOx emissions [kg]	Gasif.	FICFB	$\dot{Q}_w^+, T_{gas}$	$2.63 \cdot 10^{-5}$ $\mathrm{kg/MJ}_{th}$	Felder and Dones (2007)
PM emissions [kg]	Gasif.	FICFB	$\dot{Q}_w^+$	$7.2 \cdot 10^{-6}$ $\mathrm{kg/MJ}_{th}$	Felder and Dones (2007)
CO_2 emissions (fossil) [kg]	Gasif.	FICFB	$\dot{Q}_w^+, T_{gas}, P_{gas}$	3.16 $\mathrm{kgCO_2}$-eq/kg-oil	Dones et al. (2007)
Zinc oxide catalyst [kg]	Gas cleaning	all	$\dot{Q}_w^+$	$2.23 \cdot 10^{-2}$ $\mathrm{kg/MJ}_{th}$	Felder and Dones (2007)
Calcium carbonate [kg]	Gas cleaning	all	$\dot{Q}_w^+$	$9.38 \cdot 10^{-4}$ $\mathrm{kg/kg}_{dry}$	Equation (3.6)
Rapeseed methyl ester [kg]	Gas cleaning	CGC	$\dot{Q}_w^+, T_{gas}, P_{gas}$	from model	Gassner and Maréchal (2009b)
Gypsum disposal [kg]	Gas cleaning	all	$\dot{Q}_w^+$	$1.95 \cdot 10^{-3}$ $\mathrm{kg/kg}_{dry}$	Equation (3.6)
Zinc oxide disposal [kg]	Gas cleaning	all	$\dot{Q}_w^+$	$4.46 \cdot 10^{-4}$ $\mathrm{kg/MJ}_{th}$	Felder (2004)
CO_2 emissions (fossil) [kg]	Gas cleaning	all	$\dot{Q}_w^+$	$4.13 \cdot 10^{-4}$ $\mathrm{kg/kg}_{dry}$	Equation (3.6)
Nickel catalyst [kg]	Methane synthesis	all	$\dot{Q}_w^+$	$1.12 \cdot 10^{-14}$ $\mathrm{kg/MJ}_{SNG}$	Felder and Dones (2007)
Aluminum oxide catalyst [kg]	Methane synthesis	all	$\dot{Q}_w^+$	$1.12 \cdot 10^{-14}$ $\mathrm{kg/MJ}_{SNG}$	Felder and Dones (2007)
Nickel disposal [kg]	Methane synthesis	all	$\dot{Q}_w^+$	$2.24 \cdot 10^{-16}$ $\mathrm{kg/MJ}_{SNG}$	Felder (2004)
Aluminum oxide disposal [kg]	Methane synthesis	all	$\dot{Q}_w^+$	$2.24 \cdot 10^{-16}$ $\mathrm{kg/MJ}_{SNG}$	Felder (2004)
CO_2 emissions (biogenic) [kg]	Gasif. & SNG upgrade	all	$\dot{Q}_w^+, x_{CH_4}, x_{d,memb}$	from model	Gassner and Maréchal (2009b)
Avoided fossil NG extraction [$\mathrm{Nm^3}$]	-	all	$\dot{Q}_w^+, x_d$	from model	Gassner and Maréchal (2009b)
Avoided CO_2 emissions (fossil) [kg]	-	all	$\dot{Q}_w^+, x_d$	from model	Gassner and Maréchal (2009b)
Consumed or avoided electricity [kWh_e]	-	all	$\dot{Q}_w^+, x_d$	from model	Gassner and Maréchal (2009b)
Transport of auxiliary materials and waste [tkm]	-	all	$\dot{Q}_w^+, x_d$	30 km	distance assumed

Table 3.4 Summary of the process equipment included in the LCA model for SNG production, and of the parameters used for their scaling and adaptation to the configuration.

Name of LCI element	Process stage	Technology	Functional parameters	Source
Air dryers (reactors)	Drying	Air drying	$\dot{Q}_w^+, \omega, T_{dr}$	Gerber et al. (2011a)
Steam dryers (reactors)	Drying	Steam drying	$\dot{Q}_w^+, \omega, T_{dr}$	Gerber et al. (2011a)
Pyrolysis reactor	Biomass pretreatment	Pyrolysis	$\dot{Q}_w^+, T_{pyr}$	Gerber et al. (2011a)
Torrefaction reactor	Biomass pretreatment	Torrefaction	$\dot{Q}_w^+, T_{tor}$	Gerber et al. (2011a)
Gasification chambers (reactors)	Gasification	FICFB, CFB	$\dot{Q}_w^+, T_{gas}, P_{gas}$	Gerber et al. (2011a)
Combustion chambers (reactors)	Gasification	FICFB	$\dot{Q}_w^+, T_{gas}, P_{gas}$	Gerber et al. (2011a)
Water feed pump	Gasification	FICFB	$\dot{Q}_w^+, T_{gas}, P_{gas}$	Gerber et al. (2011a)
Oxygen compressor	Gasification	CFB	$\dot{Q}_w^+, T_{gas}, P_{gas}$	Gerber et al. (2011a)
Filter for gas cleaning	Gas cleaning	CGC, HGC	$\dot{Q}_w^+, T_{gas}, P_{gas}$	Gerber et al. (2011a)
Guard beds (reactors)	Gas cleaning	CGC, HGC	$\dot{Q}_w^+, T_{gas}, P_{gas}$	Gerber et al. (2011a)
Methanation reactors	Methane synthesis	–	$\dot{Q}_w^+, T_{meth,in}, T_{meth,out}, p_{meth}$	Gerber et al. (2011a)
Methanation pump	Methane synthesis	–	$\dot{Q}_w^+, T_{meth,in}, T_{meth,out}, p_{meth}$	Gerber et al. (2011a)
Compressors for SNG upgrade	SNG upgrade	Selexol adsorption, PSA, Membranes	$\dot{Q}_w^+, x_{CH_4}, x_{d,memb}$	Gerber et al. (2011a)
Turbines for SNG upgrade	SNG upgrade	Selexol adsorption, PSA, Membranes	$\dot{Q}_w^+, x_{CH_4}, x_{d,memb}$	Gerber et al. (2011a)
Selexol adsorbers (reactors)	SNG upgrade	Selexol adsorption	$\dot{Q}_w^+, x_{CH_4}$	Gerber et al. (2011a)
PSA absorbers (reactors)	SNG upgrade	PSA	$\dot{Q}_w^+, x_{CH_4}$	Gerber et al. (2011a)
Membranes	SNG upgrade	Membranes	$\dot{Q}_w^+, x_{d,memb}$	Gerber et al. (2011a)
Steam boiler	Steam network	all	$\dot{Q}_w^+, x_d$	Gerber et al. (2011a)
Steam turbine	Steam network	all	$\dot{Q}_w^+, x_d$	Gerber et al. (2011a)
Heat exchangers	-	all	$\dot{Q}_w^+, x_d$	Gerber et al. (2011a)

Auxiliary flows for reactors. The flows related to the process operation that are auxiliary materials consumed by the reactors – i.e. olivine, charcoal, calcium carbonate, starting oil, resulting solid waste – are assumed to be directly linked to the size of the reactors and can be expressed as:

$$\dot{m}_{j,reac} = V_{reac}(x_d) \cdot \dot{v}_{j,reac} \tag{3.4}$$

where $\dot{m}_{j,reac}$ is the mass flow rate of the LCI element j, V_{reac} is the volume of the reactor expressed in m^3, function of the decision variables x_d and $\dot{v}_{j,reac}$ is here the specific consumption rate of the material j per unit of reactor volume and time, in kg/m^3/s. The latter is assumed to be a constant and calculated from the data available for the pilot-scale process in Hofbauer and Rauch (2001), Felder (2004) and Felder and Dones (2007).

Rapeseed Methyl Ester. The quantity of required Rapeseed Methyl Ester (RME) for the scrubbing is calculated by the thermo-economic model and directly taken from its results (Gassner and Maréchal (2009b)).

Catalysts. The quantities of catalysts required for gas cleaning and methanation – i.e. zinc oxide, nickel, aluminum oxide, resulting solid waste – are assumed to be proportional to the quantity of produced SNG:

$$\dot{m}_{j,cat} = \dot{Q}^-_{SNG} \cdot \dot{v}_{j,cat} \tag{3.5}$$

where $\dot{m}_{j,cat}$ is the mass flow rate of the catalyst j, $\dot{Q}^-_{SNG}$ is the produced quantity of SNG, expressed in energy output, and $\dot{v}_{j,cat}$ is the quantity of catalyst required per unit of produced SNG. For the recycling of metal catalysts, the assumed recycling ratio is 98% (Felder (2004)), meaning that 2% of the initial mass of catalysts is considered for waste disposal.

Sulphur treatment. The chemical treatment of sulphur in the gas cleaning unit uses calcium carbonate, and produces gypsum and carbon dioxide:

$$CaCO_3 + SO_2 + \frac{1}{2}O_2 + 2H_2O \longrightarrow CaSO_4 \cdot 2H_2O + CO_2 \tag{3.6}$$

Assuming a stoichiometric ratio between the different reactants, this equation is used to calculate the required quantities of calcium carbonate for gas treatment, and the corresponding produced quantities of gypsum to be disposed, as well as the fossil CO$_2$ emissions from the gas cleaning unit. The average sulphur content of the biomass is given in Gassner (2010), and a full conversion of sulphur into sulphur dioxide is assumed.

Emissions from combustion. Nitrogen oxides (NOx) from the combustion during indirect gasification (FICFB) have been measured to $em_{NOx} = 2.63 \cdot 10^{-5}$ kg/MJ of input wood with a combustion temperature of $1123°$K and 1 bar (Felder (2004)). However, an increased combustion temperature, linked to an increased air excess, increases NOx production (Wark et al. (1998)). This effect has to be accounted for when varying the gasification temperature. Values for NOx emissions at different temperatures were available for combustion in a fluidized bed in Mann et al. (1992) and DeDiego et al. (1996) and were used to model the NOx emissions as a function of the combustion temperature, expressed as:

$$\dot{m}_{NOx} = em_{NOx} \cdot \dot{Q}_w^+ + \dot{\gamma}_{NOx} \cdot (T_g - 1123) \tag{3.7}$$

where em_{NOx} are the NOx specific emissions at $1123°$K in kg/kJ of input wood, $\dot{\gamma}_{NOx}$ is the coefficient of variation of NOx in function of the temperature, in kg/$°$K/s, estimated from Mann et al. (1992) and DeDiego et al. (1996), and T_g is the gasification temperature in $°$K. This model should be taken only as an estimation of the NOx production as a function of the temperature, and not as a detailed prediction, for which more sophisticated models would be required.

Particulate matter (PM) is as well produced by the combustion unit during indirect gasification. PM emissions have been measured to $em_{PM} = 7.2 \cdot 10^{-10}$ kg/MJ of input wood (Felder (2004)). It is assumed that PM emissions are proportional to the process scale, and are calculated by:

$$\dot{m}_{PM} = em_{PM} \cdot \dot{Q}_w^+ \tag{3.8}$$

where em_{PM} are the specific emissions of particulate matter per unit of wood, in kg/kJ. A medium size of particles, between 2.5 and 10 μm, is assumed.

In the case of direct gasification (CFB), NOx is not emitted, since this technology involves the direct use of pure oxygen, which is produced by cryonics. Its required quantity is calculated using the thermo-economic model. PM is as well not emitted in the case of direct gasification, since this technology has no direct emissions into the atmosphere.

CO$_2$ emissions. CO$_2$ is emitted at different stages of the process, and has to be separated in biogenic and fossil emissions. Indeed, biogenic CO$_2$ is neutral in terms of global warming potential, since it is part of the biologic carbon cycle and is re-absorbed by newly growing biomass after being released to the atmosphere, making the assumption of a sustainable forest management. Therefore, its weighting factor for global warming potential expressed as CO$_2$-equivalent is equal to zero, while fossil CO$_2$ has an associated weighting factor equal to one. Biogenic CO$_2$ emissions from SNG production include the emissions from the SNG upgrade, which separates the methane from the CO_2, and as well from the combustion when FICFB is used as the gasification technology. Fossil CO_2 emissions come from the periodic use of fossil light fuel oil to start the gasifier, and from the use of calcium carbonate during gas cleaning for the sulphur dioxide neutralization, which quantities can be calculated by Equation (3.6).

Logistics for auxiliary flows. The transport of the auxiliary materials and waste disposal during the process operation is as well included in the LCI. An average distance of 30 km is assumed for all auxiliary materials and waste.

Life Cycle Impact Assessment

In the present example, two different impact assessment methods are selected from the LCI database and implemented in the computational framework. These are the Ecoindicator99-(h,a) (Goedkoop and Spriensma (2000)) and the Ecoscarcity06 (Brand et al. (1998)). The first one, already briefly presented in previous chapters, is a damage-oriented approach and measures the impact on three endpoint categories, namely the human health, the ecosystem quality and the resources, that are then aggregated into a final single score. Of the three available weighting sets for the method – i.e. hierarchist, egalitarian, individualist – the hierarchist is used here, since it is commonly assumed to represent the scientific vision in the LCA community. The second impact assessment method is based on the scientifically supported goals of the Swiss environmental policy with respect to the emissions of polluting substances in air, water and soil, to the usage of resources, and to the generated waste. No impact assessment method considering only the single issue of greenhouse gases emissions is used. Indeed, these emissions are included and weighted in the two chosen impact assessment methods. This is justified by the fact that sustainable biofuels refer not only to the reduction of greenhouse gases emissions with respect to the fossil fuels. It should also be ensured that their production and use does not result in an increase of other types of environmental impacts when compared with the fossil fuels.

3.3   Comparison with conventional LCA

3.3.1   Scenarios

A scenario using the raw LCI data from Felder and Dones (2007) is adapted to the system boundaries of the present study to account for fossil natural gas substitution. The LCIA results of this reference scenario correspond to a conventional LCA of an average lab or pilot-scale process. They are then compared at a scale of 8 MW_{th}, in terms of thermal capacity of input wood, with the LCIA results for a conservative industrial process design – base case scenario – which has been developed with the thermo-economic process model as described in Gassner and Maréchal (2009a). This scenario is evaluated with and without Combined Heat and Power (CHP) production from excess heat by a steam Rankine cycle. This allows for comparing the impact of the by-products and of the process integration.

3.3.2   Results and discussion

The results are compared in bar charts where, on the upper figure, the overall impacts are presented in relative percentage using as reference the results of the conventional LCA. On the bottom figure, the impact indicator is reported and expressed in UBP[1] for Ecoscarcity06 and in points for Ecoindicator99-(h,a). On the left columns, harmful impacts of the process considering the life cycle are reported, while the right columns display the beneficial impacts that are avoided when substituting the services of the process.

Ecoscarcity06

Figure 3.3 compares the impact assessment of the conventional LCA with the developed methodology using the Ecoscarcity06 impact assessment method with detailed process contributions.

The relative differences without and with CHP are significant. The scenario described by the conventional LCA results in an overall harmful impact, even if the substitution for natural gas is accounted for. For the base case scenario without a Rankine cycle, the impact reduction is of 110% compared to the reference, which is the conventional LCA. For the base case scenario with a Rankine cycle, the reduction is of 490%. These important reductions are due to both positive and negative impacts that compensate each other. Beneficial contributions are due to the substitution of natural gas by biogenic SNG and to the electricity cogeneration for the base case scenario with a Rankine cycle.

The differences between the conventional LCA and the base case scenario without and with a Rankine cycle are mainly due to the improved conversion efficiency obtained by applying process design method, which has a direct influence on the quantity of SNG produced per unit of biomass. In the base case scenarios, the efficiency is considerably higher because the heat and power recovery is optimized in the process design using the energy integration technique. As a direct consequence, the environmental impact of the process is lower when compared to a conventional LCA based on a pilot plant design as published in the literature. For the industrial base case scenario without a Rankine cycle, the increase in efficiency more than compensates the increased harmful impacts of the equipment required to increase the efficiency of the original design. The comparison of the base case scenarios with and without electricity cogeneration further highlights the environmental benefit of a proper process

---

[1]"Umweltbelastungspunkte", in German, meaning "environmental load points"

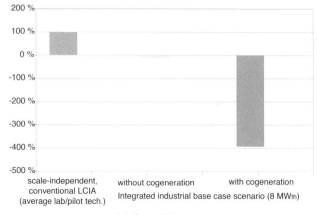

(a) Overall impacts.

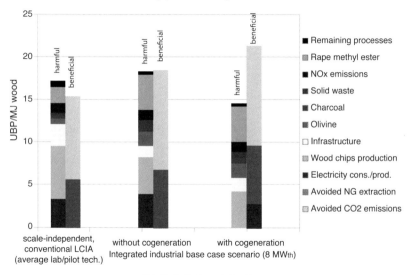

(b) Detailed contribution.

Figure 3.3 Comparison of LCIA results obtained by conventional LCA and the proposed methodology for Ecoscarcity06, single score, with detailed process contributions (adapted from Gerber et al. (2011a)).

integration. When compared with the conventional LCA, the Ecoscarcity06 impact is decreased by more than 300%.

Apart from this considerable benefit obtained from an improved conversion efficiency and regarding other contributions, it can be seen that the impact due to the wood chips production is higher for the conventional LCA. This is due to a different evaluation of the transport of wood chips in the developed methodology. While Felder and Dones (2007) use a constant transport distance, here GIS-based data are used to evaluate the average transport distance from the forest to the SNG plant for a given process scale at a given location (Stucki et al. (2010)). The issues linked with the plant location and its influence on biomass availability and transport distance have been since addressed later in Steubing et al. (2011a, 2014). Another important difference is the contribution of the infrastructure, which is higher in the case of conventional LCA. This is also due to the different approaches. While Felder and Dones (2007) scaled down linearly the total infrastructure of a methanol plant, the developed methodology offers a more detailed impact scaling method that differentiates each piece of process equipment, sized according to the process design.

Other effects of the process design on the environmental impacts are the calculation of the auxiliary flows associated with the reactors sizes – i.e. olivine, charcoal and solid waste – and of the rapeseed methyl ester (RME) consumed in the producer gas cleaning unit. These flows have a higher contribution to the environmental impact for the base case scenarios than for the conventional LCA, which is however compensated by the increase of the process efficiency. The reported differences can not be explained by the assumptions made in the LCI but are explained by the LCA models as proposed here.

Ecoindicator99-(h,a)

Figure 3.4 compares the impact assessment by conventional LCA with the developed methodology for the Ecoindicator99-(h,a) impact method with detailed process contributions. Impact is set as negative to avoid misinterpretation of the results, since the overall impact of both studied scenarios and of the conventional LCA results in a value smaller than zero, meaning therefore that all considered scenarios have a beneficial environmental impact.

Though less than with Ecoscarcity06, the relative difference between the conventional LCA and the base case scenarios is important. The reduction varies between 70 and 80% from the conventional LCA to the base case scenarios. Figure 3.4 shows that this is due to the higher process efficiency, leading to a higher SNG production per unit of biomass and therefore to a higher natural gas substitution for the base case scenario than for the conventional LCA. Unlike for Ecoscarcity06, the differences between the scenarios with and without a Rankine cycle for CHP are around 10% and thus smaller. This is due to the lower weighting in the impact assessment method of the electricity consumption or production, which has therefore a much smaller contribution with Ecoindicator99-(h,a) than with Ecoscarcity06.

3.4   Environomic optimal configurations

As outlined in the introduction, one of the main interests of linking process design and LCA is the possibility to consider environmental criteria in the technology choices at an early stage of the design and to integrate them in the optimization procedure.

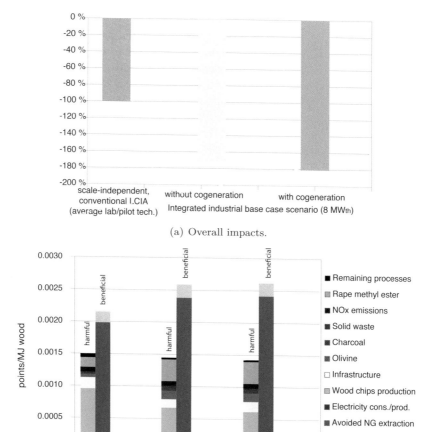

(a) Overall impacts.

(b) Detailed contribution.

Figure 3.4 Comparison of LCIA results obtained by conventional LCA and the proposed methodology for Ecoindicator99-(h,a), single score, with detailed process contributions (adapted from Gerber et al. (2011a)).

In order to study the influence of the developed LCA methodology on the optimal process configurations, a multi-objective optimization (MOO) is conducted for a selection of candidate technologies for the conversion of lignocellulosic biomass to SNG and electricity.

Thus, the trade-offs between economic, thermodynamic and environmental objectives can be calculated.

Furthermore, the effects of process scale, process design and configuration, and of the technology choice on the impacts can be highlighted.

3.4.1 Multi-objective optimization problem

Two independent optimization objectives are chosen:

1) As the economic objective, the biomass profitability $C_{profitability}$ in €/MWh$_{th}$, defined as the net profit obtained from the resource conversion, is used. In analogy to the choice of the FU for the environmental impacts, this indicator relates best to the function of the process, which is to convert a resource into useful energy services, and not only in a single product, since it includes the profits from both SNG and electricity sales:

$$C_{profitability} = \epsilon_{SNG} \cdot C_{SNG} + \epsilon_{el} \cdot C_{el} - C_w$$

$$- \frac{\Delta h^0_{RME} \dot{m}^+_{RME} \cdot C_{RME} + \dot{m}^+_{O2} \cdot C_{O2}}{\Delta h^0_w \dot{m}^+_{w,daf}}$$

$$- \frac{C_{salaries} + 0.05 \cdot C_{GR} + \frac{(1+ir)^n - 1}{ir(1+ir)^n} \cdot C_{GR}}{t_a \cdot \Delta h^0_w \dot{m}^+_{w,daf}}$$

$$(3.9)$$

with:

$$\epsilon_{SNG} = \frac{\Delta h^0_{SNG} \dot{m}^-_{SNG}}{\Delta h^0_w \dot{m}^+_{w,daf}}$$

$$\epsilon_{el} = \frac{\dot{E}^-}{\Delta h^0_w \dot{m}^+_{w,daf}}$$

C_{SNG}, C_{el}, C_{RME} and C_w corresponding to the prices of SNG, electricity, RME and biomass in € /MWh, C_{O2} to the price of oxygen in € /kg, $C_{salaries}$ to the total yearly salaries of the employees in € /year and C_{GR} to the investment cost from grass roots in € as calculated by the economic model of Gassner and Maréchal (2009b). ϵ_{SNG} and ϵ_{el} represent the product yields of SNG and electricity, Δh^0 the lower heating value on dry, ash free (daf) basis in MJ/kg$_{daf}$ and $\dot{m}$ the mass flow in kg/s of streams that enter ($^+$) or leave ($^-$) the system. The maintenance cost is assumed to amount to 5% of the investment per year, and the annualised investment is discounted at an interest rate ir over the economic lifetime n in yr of the plant. As discussed by Gassner (2010), $C_{profitability}$ – or an alternative formulation of a break-even cost for biomass in which the substraction of C_w is omitted – is the most coherent indicator to evaluate the thermo-economic design objective in a polygeneration context since it accounts for the value of all products in the same way, which would not be the case if the production costs of SNG were used. All assumed prices and parameters for the evaluation of Equation (3.9) are summarized in Table 3.5.

2) As the environmental objective, the single score of the cumulated environmental impacts described in Equation (2.4) is chosen. To highlight the importance of the choice of the environmental objective function, the optimization is performed using once the single score of Ecoscarcity06, and once the single score of Ecoindicator99-(h,a).

Table 3.5 Assumptions for process economics (adapted from Gerber et al. (2011a)).

Parameter		Value
Interest rate	ir	6%
Discount period	n	15 years
Plant availability		90%
Operators[a]		4[b] per shift
Operator salary		60 000 € per year
Maintenance costs		5% of C_{GR} per year
Wood price	C_w	33 € MWh^{-1}
Biodiesel price	C_{RME}	105 € /MWh
Electricity price (green)	C_{el}	180 € /MWh
Oxygen price	C_{O2}	(Kirschner, 2009)
SNG price	C_{SNG}	120 € /MWh
Exchange rate		1.5 CHF/€ Gassner (2010)

[a] Full time operation requires three shifts per day. With a working time of five days per week and 48 weeks per year, one operator per shift corresponds to 4.56 employees.

[b] For a plant size of 20 MW$_{th,wood}$. For other production scales, an exponent of 0.7 with respect to plant capacity is used.

Table 3.6 displays the decision variables x_d and their associated validity ranges.

One biomass pretreatment, three gasification and two gas cleaning technologies are proposed as discrete decision variables. They represent different development stages of a mature commercial technology for SNG production from lignocellulosic biomass. Indirect gasification at atmospheric pressure (FICFB) could represent a first generation of a commercial technology, while pressurized indirect gasification (pFICFB) and direct gasification (CFB) would be the second and third generations, respectively. The selected SNG upgrade technology is the membranes for all cases. The other decision variables relate to the process operating conditions.

3.4.2   Results and discussion

The Pareto curves resulting from the multi-objective optimization are shown in Figures 3.5(a) and 3.5(b). Six clusters representing different technological options of the process superstructure are displayed in different marker types. The process scale associated with the optimal process configurations is as well displayed using a color gradient. The exemplary configurations that are discussed in the subsections below are identified by a red circle and a number.

These figures clearly highlight a trade-off between the environmental impacts and the biomass profitability. This is mainly due to the effect of the process

Table 3.6 Decisions variables and associated ranges used for the multi-objective optimization (adapted from Gerber et al. (2011a)). See Gassner and Maréchal (2009b) for a detailed definition of the variables.

Name	Integer	Range	Unit
Biomass pretreatment	yes	torrefaction	–
Gasification technology	yes	FICFB, pFICFB, CFB	–
Gas cleaning technology	yes	cold, hot [a]	–
Thermal capacity as input wood (size) $\dot{Q}_w^+$	no	[5;200]	MW$_{th}$
Air dryer inlet temperature T_{dr}	no	[453;513]	K
Wood humidity at gasifier inlet ω	no	[0.1;0.3]	–
Gasification pressure P_{gas} [b]	no	[1;30]	bar
Methanation pressure P_{meth}	no	[1;30]	bar
Methanation inlet temperature $T_{meth,in}$	no	[573;673]	K
Methanation outlet temperature $T_{meth,out}$	no	[573;673]	K
Wobbe index after CO_2 removal	no	[13;13.8]	kWh/Nm^3
High (retentate) pressure of membrane 1	no	[5;50]	bar
High (retentate) pressure of membrane 2	no	[5;50]	bar
High (retentate) pressure of membrane 3	no	[5;50]	bar
Molar stage cut of stage 2	no	[0.2;0.6]	–
Molar stage cut of stage 3	no	[0.2;0.6]	–
Remainder of purification inlet to stage 2	no	[0;1]	–
Steam production pressure	no	[40;120]	bar
Steam superheat temperature	no	[623;823]	K
Number of steam utilization levels (integer)	no	[0;3]	–
Steam bleeding temperature	no	[323;523]	K
Steam condensation temperature	no	[293;383]	K
Minimum temperature difference in HEX network	no	[1;2]	K

[a] applies only to CFB technology
[b] applies only to pFICFB and CFB technologies

scale, since an increase in the size leads simultaneously to an increase of the environmental impacts but also to an increase of the biomass profitability for all clusters and both environmental objectives.

A second important aspect is the effect of the technology evolution, represented by the different clusters, since switching to a more advanced technology as it becomes available at a commercial scale tends to simultaneously increase the biomass profitability and decrease the environmental impacts.

A third aspect is the effect of the choice of the environmental objective, since with Ecoscarcity06 the FICFB configuration using a pretreatment by torrefaction performs less good than the FICFB configuration without it, while this is the opposite in the case of Ecoindicator99-(h,a). These three aspects are

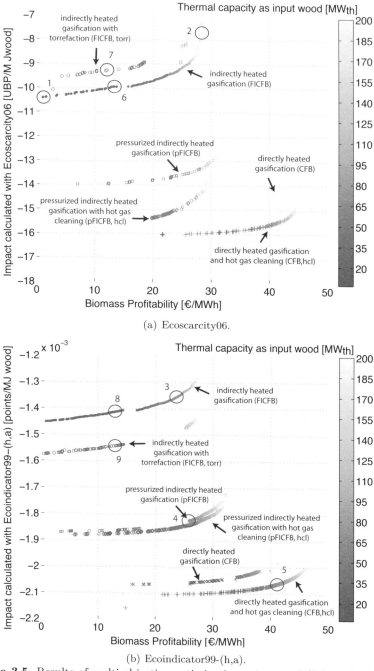

(a) Ecoscarcity06.

(b) Ecoindicator99-(h,a).

Figure 3.5 Results of multi-objective optimization using one LCIA method as the environmental objective, biomass profitability as the economic objective, at multiple scale. (a) Ecoscarcity06 as the environmental objective and (b) Ecoindicator99-(h,a) as the environmental objective (adapted from Gerber et al. (2011a,b)).

discussed more in details and illustrated by exemplary configurations in the paragraphs below. The detailed characteristics and operating conditions of each exemplary configuration are available in Annex B.

Process scale

The increase of the profitability with process size is mainly due to the economies of scale. Although the efficiency in SNG production is increasing and the contribution of infrastructure to the environmental impact is decreasing with the scale, this effect is compensated by an increase of the contribution of biomass logistics, by an increase of the auxiliary materials, such as olivine and charcoal, and by a decrease of the electricity produced.

To illustrate this, the specific contributions are displayed for two points of the FICFB cluster, one at 5 MW$_{th}$ (configuration 1 on Figure 3.5), and one at 200 MW$_{th}$ (configuration 2 on Figure 3.5) in Figure 3.6 using the Ecoscarcity06 indicator. The total difference between the two configurations is an increase of 26% in the total impact when the process scale increases from 5 to 200 MW$_{th}$.

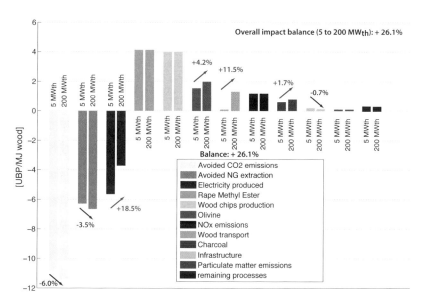

Figure 3.6 Impact contribution for configurations 1 and 2 of Figure 3.5 using FICFB at different scales. Important relative differences between contributions are indicated using the 5 MW$_{th}$ as the reference (adapted from Gerber et al. (2011a)).

The overall benefit due to the increase of SNG production and to a lesser extent to the decrease of the specific contribution of infrastructure is of 10%. However, these benefits are more than compensated by the cumulated increase of the impacts mostly due to electricity production and biomass logistics, and to a lesser extent to the increase of the specific consumption in auxiliary materials, the global increase being of 36%.

Both the decrease of the contributions for infrastructure and the increase for auxiliary materials required for gasification are a direct consequence of the adopted methodology that links the LCI flows to the process flowsheet. More precisely, the latter is an effect of the gasifier scaling, whose specific volume per unit of FU increases with size. This has a direct effect on the trade-off between economic and environmental aspects, and could not have been highlighted with a conventional LCA methodology.

For almost all the Pareto curves of Figure 3.5, up to a certain process scale, the impact tends to increase slightly while the biomass profitability increases in an important way, due to the economies of scale. This threshold process scale varies between around 50-120 MW_{th}, depending on the technology choice and on the chosen environmental objective function. Beyond this process scale, the biomass profitability tends to reach a certain value without increasing anymore, and the impact increases in an important way due to the biomass logistics and the auxiliary materials, the latter being more important in the case of atmospheric FICFB technologies as it is discussed in the subsection below.

Since the trade-off between environmental impacts and biomass profitability is mainly due to the process scale, the question of the optimal plant size arises. This is however controlled not only by the conversion stage, but as well by the SNG plant location. Indeed, the impacts due to the wood harvesting and logistics are linked with the plant location, which influences the geographical area needed to supply a plant from a given size. Therefore, geo-localized models, based on GIS, are required for these two stages. The multi-scale integration of GIS-based models for the wood availability, harvesting, logistics and of non-site specific SNG conversion technologies and substitution of energy services has been realized in Steubing et al. (2011a, 2014). It is demonstrated that the optimal economic SNG plant sizes are between 100 and 200 MW_{th}, while the environmental ones are between 5 and 40 MW_{th}. This trade-off can however be minimized for plants above 25 MW_{th}.

Technology evolution

Another advantage of the developed methodology is the possibility of analyzing both the economics and the environmental impact of technology evolution. The effects of three technological evolutions are illustrated by displaying the contributions for three points at a fixed scale of 60 MW_{th} for FICFB (configuration 3 on Figure 3.5), for pFICFB (configuration 4 on Figure 3.5) and for CFB with hot gas cleaning (configuration 5 on Figure 3.5) in Figure 3.7, using the Ecoindicator99-(h,a).

Pressurizing the indirectly heated gasification technology would allow for an important impact reduction due to a more compact gasifier design, the global reduction being in this example of 35%, despite a slightly decreased SNG production. The volume reduction decreases consequently the amount of auxiliary

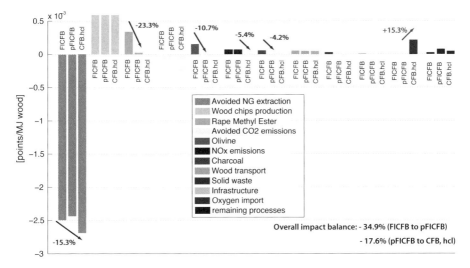

Figure 3.7 Impact contribution for Ecoindicator99-(h,a) at 60 MW$_{th}$ for three configurations using different technologies (configurations 3, 4 and 5 of Figure 3.5). Important relative differences between contributions are indicated using the FICFB as the reference (adapted from Gerber et al. (2011a)).

materials required for gasification, olivine and charcoal representing together 15% of impact reduction. Moreover, since pFICFB operates under pressurized conditions, the volume of gases coming from the gasifier and its associated amount of RME required for gas cleaning is drastically reduced, and decreases the impact by 23%. Switching from indirect gasification technology to direct, steam-blown gasification (CFB) with hot gas cleaning reduces the impact even further – 53% when compared with the base case of atmospheric FICFB – as CFB gasifier does not release any combustion gases to the atmosphere. The nitrogen oxides impacts are thus removed. This last technological option presents as well an increased efficiency and produces therefore more SNG per unit of wood, although this is however compensated by the cryogenic oxygen import contribution.

Choice of environmental objective function

A last aspect highlighting the influence of the integration of environmental impacts in the optimization procedure is the influence of the environmental objective in the optimal process design. While the Pareto curves show similar ranges of biomass profitability for both objective functions, the actual optimal process configurations change depending on the choice of the environmental objective function. Moreover, for the two scenarios using FICFB, in the case of Ecoscarcity06 the use of a torrefaction pretreatment decreases the environmental performance while it increases it in the case of Ecoindicator99-(h,a). To illustrate the effects of the choice of the environmental objective function on the optimal process design, four points at the same scale using different environmental objective functions and with or without a torrefaction pretreatment are compared. Their detailed characteristics are shown in Table 3.7. The impact contributions of these two points are displayed in Figures 3.8(a) and 3.8(b).

Table 3.7 Comparison of four optimal process configurations for different environmental objective functions and technology choices (configurations 6, 7, 8 and 9 of Figure 3.5).

	FICFB - Ecoscarcity 06	FICFB - Ecoindicator 99-(h,a)	FICFB,tor - Ecoscarcity 06	FICFB,tor - Ecoindicator 99-(h,a)
Gasification technology	FICFB	FICFB	FICFB	FICFB
Biomass pretreatment	none	none	Torrefaction	Torrefaction
Process scale [MW$_{th}$]	12.7	12.3	12.9	12.1
Profitability [€/MWh]	3.7	13.8	11.8	13.4
Total electricity production [MW$_e$]	0.761	0.015	0.444	−0.18
Relative electricity production [MW$_e$/MW$_{th}$]	0.06	0.001	0.034	−0.015
Total SNG production [MW]	7.9	8.6	8.4	8.7
Relative SNG production [MW/MW$_{th}$]	0.620	0.703	0.648	0.703

The optimal process configurations using Ecoscarcity06 favors an increased electricity production and a reduced SNG production, while Ecoindicator99-(h,a) favors clearly the SNG production, due to the different weightings of the substituted services in the two impact assessment methods. This leads to different process configurations for the same clusters depending on the selected environmental objective. Using a torrefaction unit for the biomass pretreatment decreases the electricity production but increases the SNG output, which explains why this configuration has a less good environmental performance in the case of Ecoscarcity06 and a better one in the case of Ecoindicator99-(h,a).

This example shows how the integration of environmental impacts in the optimization procedure influences the engineering decisions related to the final process design. One could argue that in such a case where two environmental objectives functions lead to different designs, the final decision should be taken therefore on an economic basis. This is the case for the two configurations using FICFB with a torrefaction pretreatment, where the optimal configuration with Ecoindicator99-(h,a) has a clearly higher biomass profitability than the one with Ecoscarcity06 (see Table 3.7).

However, different configurations can have a similar economic performance, like this is the case for the two optimal configurations using FICFB without a biomass pretreatment, which both have a very similar biomass profitability (see Table 3.7). In such a case where economics is not the major decision criterion, the design decisions might be taken on an environmental basis.

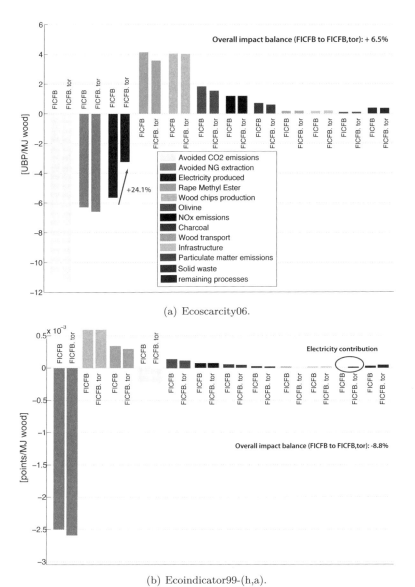

(a) Ecoscarcity06.

(b) Ecoindicator99-(h,a).

Figure 3.8 Impact contribution for a 12 MW$_{th}$ configuration using FICFB with or without a torrefaction pretreatment (configurations 6, 7, 8 and 9 of Figure 3.5). (a) Ecoscarcity06 as the environmental objective and (b) Ecoindicator99-(h,a) as the environmental objective. Important relative differences between contributions are indicated using the FICFB without a pretreatment as the reference (adapted from Gerber et al. (2011a,b)).

In addition, this issue of the choice of the environmental objective function illustrates the importance of selecting the appropriate substitution scenario for energy services.

3.4.3   Substitution in multi-product systems

Because of the importance of the substitution of energy services in the environmental impacts, the question arises whether the reduction in environmental impacts is strictly correlated with the increase in efficiency. To address this issue, the optimization strategy can be changed by calculating this time the trade-off between the environmental objective function and the production of both energy services. In order to remove the effects due to the scaling on the environmental impacts, the calculations are made for a 20 MW_{th} fixed process size, the other decision variables remaining unchanged. The multi-objective optimization is using three objectives, one of them being the environmental impact, the two others being the specific SNG production and the electricity yields to be maximized. Multi-objective optimization is performed twice with respect to both Ecoscarcity06 and Ecoindicator99-(h,a) as the environmental objective to be minimized. The results are displayed in Figures 3.9(a) and 3.9(b), for the FICFB cluster. In these figures, the thermodynamically optimal process configurations are displayed as well. It corresponds to the flowsheets that maximize the overall "chemical" efficiency, in which the contributions of SNG and electricity are weighted by the conversion of SNG to electricity in a state-of-the-art natural gas combined cycle with an efficiency of 55%, as explained in Gassner (2010).

The results present an opposite trade-off between the two energy services. While Ecoscarcity06 favors the electricity production, Ecoindicator99-(h,a) objective favors the SNG production. These differences are due to the different weighting of energy services in the LCIA step. The thermodynamic optimum has an overall efficiency of 73.7% in terms of SNG-equivalent, with a share of 13.59 MW of SNG and 0.65 MW_e of electricity. Due to the different weighting of SNG and electricity in the LCIA methodologies, this does not correspond to any of the environmental optima, especially in the case of Ecoscarcity06 where it is situated in the middle of the range of environmental impacts associated with optimal configurations. This result suggests therefore that in the case of a process producing multiple energy services, reducing the environmental impacts cannot be assimilated uniquely to an increase in the process efficiency.

The present example highlights as well the important issue linked with the substitution weighting, which leads to favor the electricity production for one environmental indicator and the SNG production for the other. This difference is due to the higher weighting attributed to the use of fossil resources in the Ecoindicator99-(h,a). As a consequence, Ecoindicator99-(h,a) does not favor the substitution of the Swiss mix, which has a relatively low fossil resource content. The opposite is the case for Ecoscarcity06 that attributes a higher weighting to the substitution of nuclear electricity, which represents an important share of the Swiss mix.

Another consequence is the importance of the assumption made for electricity substitution. In the present case, the Swiss mix including the imports is

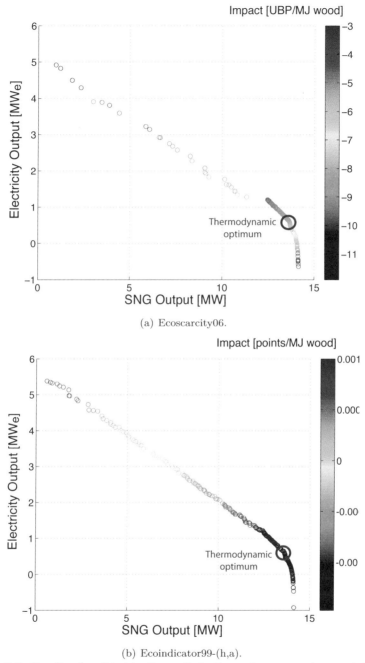

(a) Ecoscarcity06.

(b) Ecoindicator99-(h,a).

Figure 3.9 Results of multi-objective optimization using one environmental objective, SNG and electricity output as other objectives, at a fixed scale of 20 MW$_{th}$ for FICFB. (a) Ecoscarcity06 as the environmental objective and (b) Ecoindicator99-(h,a) as the environmental objective (adapted from Gerber et al. (2011a)).

substituted. The assumption made for the substitution of electricity in any LCA study is however often controversial. The results are likely to be sensitive to this assumption, especially if an electricity mix containing a higher share of fossil resources is substituted. Indeed, this would probably be a better assumption if the optimization had to be done at the present time instead of a few years ago, due to the decision of the Swiss Federal Council to abandon progressively nuclear power that happened in between. However, this difference has the benefit to highlight the influence that a political decision can potentially have on engineering decisions.

3.5   Conclusions on thermochemical wood conversion

In this chapter, the developed methodology for the integration of LCA in the conceptual design of renewable energy systems has been illustrated by its application to extend a thermo-economic model for the thermochemical conversion of lignocellulosic biomass to Synthetic Natural Gas (SNG) and electricity to an environomic model. Formulating the LCI as a function of the design variables of the thermo-economic model allows for considering the environmental performance calculated by LCIA together with the thermo-economic indicators as objective functions in the process multi-objective optimization at an early stage of the process synthesis.

The results of the comparison of the developed methodology with a conventional design and LCA highlight the influence of the process design and integration on the environmental impacts calculated with the LCA. Indeed, the effect of the increase in efficiency due to a proper process integration that accounts for an optimal heat and power recovery leads to a particularly high impact reduction when compared to a conventional design and LCA, for which such aspects are not considered.

The application of multi-objective optimization to different technological choices and evolution stages has allowed for calculating and analyzing the trade-offs between environmental, economic and thermodynamic objectives. The application of the method to the present case study has emphasized that special care needs to be taken in the choice of the functional unit and of the objective functions for polygeneration systems. Although convenient in single product systems, considering the quantity, impact and cost of the product is not a valid assumption anymore since the real function of such systems is to convert a limited resource into useful services. The results illustrate the suitability of the method to identify optimal process configurations from an environomic point of view in the decision-making procedure. The choice of the environmental objective function in the optimization is particularly important since it influences the optimal process configuration for a given economic performance. This was highlighted by the comparison of the optimal configurations using either

Ecoscarcity06 or Ecoindicator99-(h,a) as the environmental objective, for the technology of atmospheric steam-blown gasification (FICFB) with or without a torrefaction pretreatment. In this regard, the integration of environmental impacts is likely to have an important influence on the engineering decisions linked with process design.

Another important outcome of this case study is the possible non-correspondence of the thermodynamic optimum with the environmental optima. The energy service substitution and therefore the increase in energy efficiency are key points for the reduction of environmental impacts. However, the thermodynamic and environmental objectives are not strictly correlated one with the other, especially in the case of a process producing multiple energy services. Subsequently, this confirms the need for integrating the environmental dimension in the optimization procedure as a separate objective.

Finally, this application case study highlights the importance of the impact caused by the logistics, the auxiliary materials and the off-site emissions associated with the process operation. These are usually not accounted for in conventional process design considering only thermo-economic objectives. It is therefore important to create a functional link between process design decisions and indirect emissions, which can be addressed with this method but not with a conventional approach. In the present case study, this has allowed for demonstrating that the technological evolutions from the atmospheric steam-blown indirect gasification (FICFB) to more compact technologies such as pressurized indirect gasification (pFICFB), direct oxygen-blown gasification (CFB), and from cold gas cleaning to hot gas cleaning, reduce the impacts due to the auxiliary materials and emissions in an important way. Moreover, the developed methodology allows for impact scaling, which increases with process scale in the present case, due to the biomass logistics and to the specific consumption of auxiliary materials.

CHAPTER 4

Application to enhanced geothermal systems

4.1 Geothermal cogeneration systems

Geothermal energy is a renewable energy resource currently gaining interest. The International Energy Agency (IEA) expects in its roadmap a consequent increase both in electricity production and in direct use for heating from geothermal energy (IEA (2011)). One of the major advantages of geothermal energy is that it allows for providing baseload power at constant operating conditions, unlike other renewable energy sources, such as wind or sun, that are subject to climatic variations. In countries with favorable geological conditions, such as Iceland, conventional hydrothermal resources can be developed to increase the share of geothermal heat and power. However, in most of the countries without such favorable conditions, including Switzerland, the potential hydrothermal resources do not have a sufficient temperature level for electricity production. To develop geothermal combined heat and power or single power production in such countries, unconventional deeper resources with a sufficient temperature level have to be targeted for future exploitation, using the emerging technology of Enhanced Geothermal Systems (EGS).

EGS are engineered reservoirs to extract heat from low permeability or low porosity geothermal resources (Tester et al. (2006); DiPippo (2008)). They generally involve the exploitation of so-called Hot Dry Rock (HDR) resources, which have a sufficiently high temperature for power production but lack the necessary permeability. Such HDR resources can be engineered by hydraulic fracturing, which involves the injection of highly pressurized cold water through an injection well, and as a consequence creates new fractures and opens the existing ones. This repeated operation creates thus a reservoir acting like a heat exchanger heating up cold water, which is then extracted at a high temperature through one or more extraction wells. The heat available from this geofluid is

then used in a thermodynamic cycle to generate electricity and potentially heat
for direct use, if Combined Heat and Power (CHP) is considered. The principle
is illustrated in Figure 4.1.

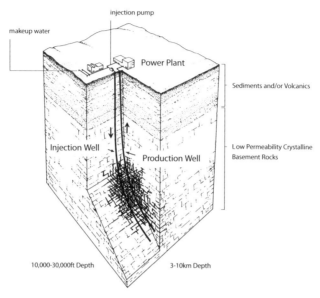

Figure 4.1 Conceptual view of a two-well EGS in a hot rock in a low-permeability
crystalline basement formation (reproduced from Tester et al. (2006) with permission).

EGS is currently still an emerging technology with a few pilot-scale projects,
notably in France (Genter et al. (2012)), Australia (Wyborn et al. (2005))
and Switzerland (Haring (2004)). In addition to the issues linked with the
geo-engineering, such as permeability enhancement or induced seismicity (Majer et al. (2007)), the economic competitiveness (IEA (2011)), the thermodynamic efficiency of the conversion system (DiPippo (2004); Kanoglu and Dincer
(2009)) and the environmental impacts (Evans et al. (2009)) are as well important criteria to be considered to determine the optimal configurations of
geothermal energy conversion systems. These concern the targeted exploitation depth, the choice of the conversion cycle and of its operating conditions
(Desideri and Bidini (1997); Hettiarachchi et al. (2007); Franco and Villani
(2009)), as well as the district heating parameters for CHP systems (Ozgener
et al. (2007); Guo et al. (2011)).

Therefore, the combined production of electricity and district heating from EGS
is an interesting application example of the developed methodology presented
in section 2.2. Due to the seasonal variation of the demand in district heating,
the multi-period aspects included in the methodology explained in subsection
2.2.6 can be illustrated. Moreover, there has been presently very few examples
of the application of LCA to the power generation from geothermal systems
(Santoyo-Castelazo et al. (2011); Frick et al. (2010)), the latter being the only

one specifically applied to EGS. Eventually, unlike for biofuels conversion processes used as a case study in the previous chapter, the life cycle inventory of a deep geothermal system includes many elements and processes occurring during the construction phase, and not only during the operation. This makes it a good case study to illustrate the full life cycle aspects. The objective of this application example is therefore to determine the promising environomic configurations for the future development of EGS in Switzerland, in terms of targeted depth, conversion technologies and share between electricity and district heating, for given geological conditions.

4.2 Models description

The following section describes the thermo-economic models used for the application case study of EGS, as well as their extension to LCA models following the methodology explained in section 2.2. The computational framework described in Figure 2.7, considering independent operation periods, is used to design the geothermal energy conversion system. The geothermal system design aims at defining, for a given geographical location, the geothermal depth, the configuration in terms of equipment sizes and operating conditions of the conversion system, as well as the operation strategy to supply the energy services of the area, considering economic, thermodynamic and environmental criteria. It is a multi-period problem that accounts for the seasonal variations of the demand in energy services. Three subsystems composing a geothermal system are considered:

1) the potential geothermal resources from which heat can be harvested. In the present application case study, only HDR is considered.

2) the potential conversion technologies.

3) the geo-localized demand profiles in energy services.

According to the methodology described in Figure 2.7, the components of the three sub-systems are modeled separately in a first time for each period. They are then integrated together to build the overall system to supply energy services using the energy integration technique. The decision variables of this MILP subproblem are the utilization rates of the different technologies of the superstructure simulated at the previous step. At the end of the single-period sequence, thermo-economic and LCA indicators of the integrated system are calculated. The whole sequence is repeated for each period. Then, overall performance indicators are calculated for the yearly operation of the system by combining the seasonal performance indicators. It includes the objective functions of the multi-objective optimization (MOO) master MINLP problem, solved using the evolutionary algorithm.

Due to the geological uncertainties, the methodology is applicable only to orientate the decision-making and the future development of geothermal energy in a given area, for which the geology is known and the demand in energy services characterized.

Here, the case study considers the construction of an EGS located on the Swiss Plateau, for which the geological conditions are known. The local demand in energy services has been as well characterized.

4.2.1   Thermo-economic model

The following subsection describes the developed thermo-economic models of the components for the three subsystems involved in the design of a geothermal system – resources, technologies and services. The currency exchange rate in all economic calculations of the present chapter is 0.92 CHF/\$ (01.12.2012).

Geothermal resources

At a given location, the exploitable geothermal resources are defined by depth, temperature and expected mass flow rates. Specifically regarding EGS, which is considered in this application case study, the model assumes a mature commercial technology, expected to be achieved in the next decades (Tester et al. (2006)), with one injection well and two extraction wells. Its conceptual flowsheet is displayed in Figure 4.2.

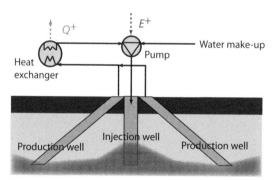

Figure 4.2 Conceptual flowsheet for the model of EGS (adapted from Gerber and Maréchal (2012a)).

Assuming a sustainable exploitation, the temperature of the HDR resource can be predicted as a function of its depth z:

$$T_{EGS,z} = T_0 + z \cdot \frac{\Delta T}{\Delta z} \tag{4.1}$$

where T_0 is the temperature at the surface, assumed to be 12°C and $\frac{\Delta T}{\Delta z}$ the geothermal gradient. The temperature difference between the hot rock and the water at the extraction wells is calculated by:

$$T_{EGS,ext} = T_{EGS,z} - \Delta T_{EGS} \tag{4.2}$$

where ΔT_{EGS} is the temperature difference between the hot rock and the water at the extraction. The thermal power available from the geothermal resource is then calculated by:

$$\dot{Q}^+_{EGS} = \dot{m}_{ext} \cdot (h(P_{ext}, T_{EGS,ext}) - h(P_{ext}, T_{EGS,inj})) \qquad (4.3)$$

where $\dot{m}_{ext}$ is the expected mass flow rate from the EGS, in kg/s, h is the specific enthalpy of the geothermal water, in kJ/kg, P_{ext} is the pressure of geothermal water at the extraction well, in bar, $T_{EGS,ext}$ is the temperature of the water at the extraction well and $T_{EGS,inj}$ is the temperature before the mixing with cold make-up water for reinjection.

For the expected mass flow rate $\dot{m}_{ext}$, the pilot EGS project of Soultz-sous-Forêts, in France, has a planned final extraction mass flow rate between 70 and 100 kg/s (Cuenot et al. (2008)). The project of Basel, in Switzerland, was targeting 100 kg/s (Haring (2004)). Tester et al. (2006) assume 80 kg/s for a mature technology. Therefore, a value of 90 kg/s for extraction is assumed here. The minimal reinjection temperature $T_{EGS,inj}$ is assumed to be 70°C and the temperature difference ΔT_{EGS} between the bedrock temperature and the geofluid at the extraction well to be 20°C, based on the data of Soultz-sous-Forêts (Cuenot et al. (2008)). The depth is considered customizable and goes from 3000m, representing the upper limit of the bedrock in Switzerland, down to 10 000m, representing the limit for the accessible resource with the current drilling technology (Tester et al. (2006)). The temperature is calculated as a function of the depth, and for the present case study a geothermal gradient $\frac{\Delta T}{\Delta z} = 0.0352°C/m$ from 3000m is used, this value being considered to be representative for the Swiss Plateau (Sprecher (2011)).

The costs for the construction of the EGS were taken from Tester et al. (2006) and updated with the inflation rate for 2012. Figure 4.3 shows the resource temperature and the drilling costs for one well as a function of the resource.

Conversion technologies

The superstructure of conversion technologies includes different CHP cycles for geothermal power generation (DiPippo (2008)):

- single-flash (1F) systems (3-90 MW$_e$)
- double-flash (2F) systems (3-90 MW$_e$)
- organic Rankine cycle (ORC) (0.25-10 MW$_e$) without or
- with an intermediate draw-off (ORC-d) at the turbine (0.25-10 MW$_e$)
- ORCs operating under supercritical conditions (ORC-s) (0.25-10 MW$_e$)
- ORCs with two evaporation levels (ORC-2) (0.25-10 MW$_e$)
- Kalina cycle based on the KCS-11 design (Mlcak (2002)) (0.25-10 MW$_e$)

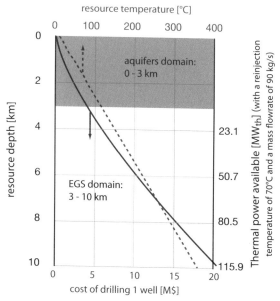

Figure 4.3 Temperature (calculated from Sprecher (2011)) and economic (calculated from Tester et al. (2006)) model of geothermal resources as a function of depth (adapted from Gerber and Maréchal (2012c)).

The validity ranges for the minimal and maximal sizes of the geothermal equipments have been as well based on the data available in DiPippo (2008) for the different conversion technologies.

Flash systems directly use the geofluid for electricity generation, as illustrated by the conceptual flowsheet of a single-flash system in Figure 4.4. The liquid is separated in a steam phase and in a liquid water phase using a flash drum. The steam is then used for electricity generation, while the liquid phase can be used for additional heat generation before being reinjected. The system can generate more electricity by a adding a second flash separation stage, as illustrated by the conceptual flowsheet of a double-flash system in Figure 4.5.

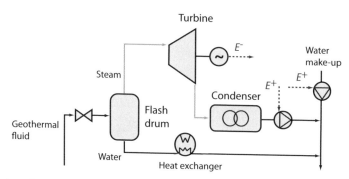

Figure 4.4 Conceptual flowsheet for the model of the single-flash (adapted from Gerber and Maréchal (2012a)).

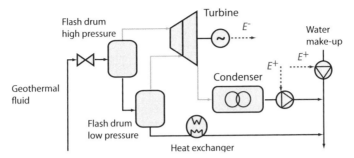

Figure 4.5 Conceptual flowsheet for the model of the double-flash (adapted from Gerber and Maréchal (2012a)).

Binary cycles, such as ORCs or Kalina cycles, use a secondary fluid for heat exchange with the geofluid and electricity generation. The secondary fluid can either be an organic fluid, in ORCs, or a mixture of water and ammonia reducing the exergy losses linked with the evaporation of the working fluid, in Kalina cycles. Several variations of the basic single-loop ORC design, illustrated in Figure 4.6, are possible.

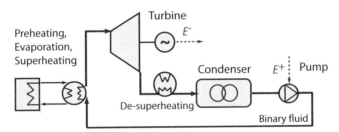

Figure 4.6 Conceptual flowsheet for the model of the single-loop ORC, working in sub-critical or supercritical conditions (adapted from Gerber and Maréchal (2012a)).

An intermediate steam draw-off can be done at the turbine, as illustrated in Figure 4.7. This allows for supplying heat at the required temperature for the district heating within the cycle.

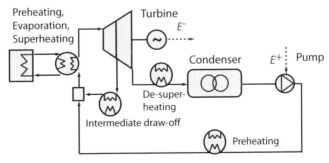

Figure 4.7 Conceptual flowsheet for the model of the ORC with an intermediate draw-off (adapted from Gerber and Maréchal (2012a)).

Another variation of the basic ORC design is to add a second evaporation stage for the working fluid, as illustrated in Figure 4.8. This reduces the exergy losses associated with the evaporation of the working fluid and allows therefore for increasing the electricity production. Another possibility to increase the power output from an ORC is to operate at supercritical conditions.

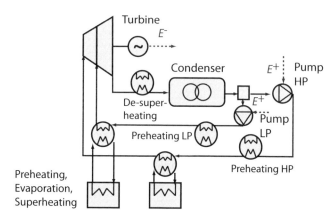

Figure 4.8 Conceptual flowsheet for the model of the ORC with two evaporation levels.

For operating the ORCs, different potential working fluids are considered. Their thermodynamic characteristics are displayed in Table 4.1.

ORCs can be used either as a single technology or as bottoming cycles in combination with flash systems. To simulate the cycles, calculate the corresponding pressures, temperatures, energy and mass flow rates, the flowsheeting software Belsim-Vali (Belsim (2011)) is used.

Table 4.1 Characteristics of the different potential working fluids included in the ORCs models, taken from Belsim database (Gerber and Maréchal (2012c)).

Fluid	Molecular Weight [kg/kmol]	Critical temperature [°C]	Critical pressure [bar]	Boiling temperature [°C] at P_{atm}
n-pentane	72.151	196.63	33.75	36.05
cyclo-butane	56.108	186.85	49.85	12.51
iso-butane	58.124	134.98	36.48	−11.83
iso-pentane	72.151	187.25	33.34	27.85
benzene	78.114	288.95	49.24	80.15
toluene	92.141	318.85	42.15	110.65
n-butane	58.124	152.01	37.97	−0.48
R134a	102.032	101.06	40.59	−26.07

An example of the simulation results for an ORC with an intermediate draw-off at the turbine using iso-butane is shown in Figure 4.9.

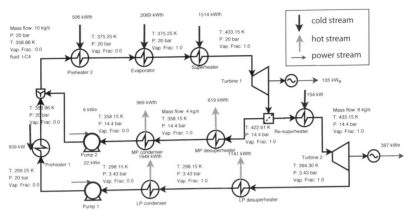

Figure 4.9 Example of simulation results for an ORC with an intermediate draw-off using iso-butane (adapted from Gerber and Maréchal (2012b)).

The heat exchanger network performances are simulated using the energy integration method, as illustrated previously in Figure 2.7. This allows for calculating the system performances without having to design *a priori* a heat exchanger network. The optimal operating conditions and flows in the system are therefore calculated and the heat exchanger design is realized once the optimal conditions are identified. The nominal heat loads, power outputs and the calculated temperature levels are used for the heat and power integration.

Finally, the results of the simulation and of the process integration are used for equipment sizing, such as turbines, pumps, heat exchangers or flash drums. Non-linear power correlations available from Turton et al. (1998) and Ulrich (1996) are used for calculating the investment cost associated with each equipment.

Energy services demand profiles

In order to represent the seasonal variations of the demand in district heating – space heating and hot water – the yearly operation is divided in 4 periods corresponding to a different average ambient temperature for the corresponding period duration. For each period and each building, a building model is used to calculate the temperature-enthalpy profile of the heat demand. The demands are then aggregated to form a demand Grand composite curve, following the methodology of Girardin et al. (2010).

Figure 4.10 presents four corresponding composite curves calculated for a residential area of the city of Nyon, located in the Swiss Plateau, with a built surface of 177 000 m^2 and a total demand of 9153 MWh$_{th}$/yr in district heating and domestic hot water.

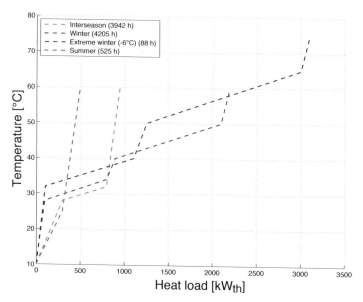

Figure 4.10 Seasonal demand profiles in space heating and hot water for a residential area in the Swiss Plateau, for a nominal installed capacity of 3.1 MW_{th} (adapted from Gerber and Maréchal (2012a)).

These composite curves concern a complete area. Therefore, it includes on the same graph the heat demands for buildings differing by age, quality and usage, as well as the hot water production. The four considered periods are summer (525 h), interseason (3942 h), winter (4205 h) and extreme winter (88 h), this last period corresponding to the nominal installed capacity of the district heating. The durations of the periods are then used to calculate the annual operating costs, since the operating conditions differ for each period.

Since one of the objectives of the case study is to calculate the optimal shares between the electricity production and the heat production, the nominal installed capacity of the district heating system is therefore considered as a decision variable, while the temperature profiles are considered as being independent of the area covered by the district heating. The electricity production is not constrained and left in any case as variable.

4.2.2   Life Cycle Assessment model

The LCA model for the different components involved in the life cycle of an EGS is established following the methodology described in Figure 2.2.

Functional Unit definition

The decision that has to be taken in the present case study concerns the configuration of the EGS, in terms of depth, conversion technology and share between district heating and electricity, in the geological and economic context of Switzerland. This has for consequence that the functional unit can not be defined in terms of heat, electricity or available geothermal heat, since these quantities all vary among the different candidate configurations. Therefore, the functional unit is here defined as the construction, operation and dismantling of one EGS, either for single electricity production or for the combined production of heat and electricity.

The substitution of produced energy services – the avoided impacts from conventional production of electricity and heating by fossil resources – has to be included, to account for the efficiency of the conversion system. An expected lifetime of 30 years is assumed for the EGS.

Life Cycle Inventory model

The three different element types involved in the LCI – flows of the process flowsheet, flows of auxiliary materials and process equipment – are identified and quantified, following the methodology described in section 2.2.3.

Figure 4.11 displays the different elements of the Life Cycle Inventory that are included in the system boundaries.

Transportation of auxiliary materials is not displayed on the figure but is included in the LCI model. The avoided impacts from the produced services are calculated by substitution of the equivalent services produced from fossil natural gas resources with the currently best available technologies. For the electricity production, a natural gas combined cycle (NGCC) with an electrical efficiency of 57% is assumed, while a natural gas boiler with an efficiency of 95% is assumed for the heat supply. In order to account for the off-site emissions, the LCI database ecoinvent® (Frischknecht et al. (2005)) is used to find equivalences for each LCI element. These are listed in Annex A.2.

According to the developed methodology, for the flows already included in the thermo-economic model, the calculated value is directly taken from it, and for the process equipment, the impact scaling method presented in section 2.2.4 is used. For the auxiliary materials, the formulation is developed on a case-by-case basis following Equations (2.8) and (2.9), depending on the life cycle stage to which the LCI element belongs.

Table 4.2, 4.3, 4.4 and 4.5 summarize all the elements included in the LCI for the construction, the operation and the end-of-life, respectively, with the parameters that allow for calculating them as a function of the geothermal system configuration, as well with their reference quantities v_j when this one is necessary. Table 4.6 gives the description of these parameters.

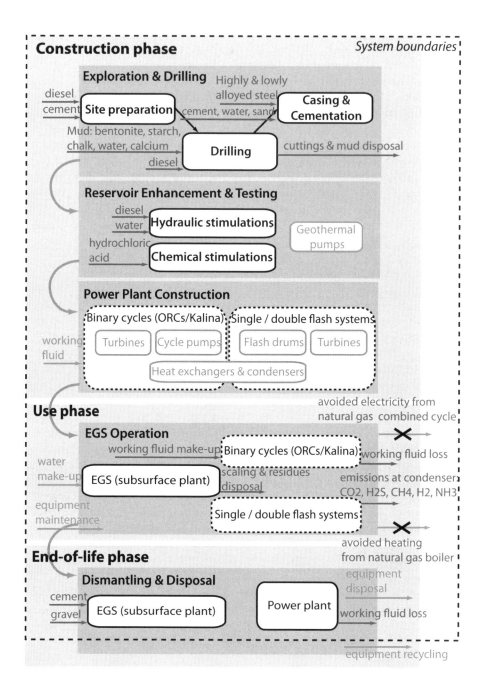

Figure 4.11 Major flows of the LCI of an EGS over its full life cycle, including flows of the thermo-economic model (green), flows of auxiliary materials (blue) and process equipment (orange) (adapted from Gerber and Maréchal (2012c)).

Table 4.2 Summary of the different LCI elements included in the LCA model for construction. The parameters and reference quantities used to perform the scaling and adapt them to the configuration are as well listed (adapted from Gerber and Maréchal (2012c)).

Name of the LCI element & Unit	Life cycle stage	Conv. tech.	Functional parameters	Reference quantity v_j	Source
Diesel for site preparation [MJ]	Site preparation	all	s	20 000 MJ/site	Frick et al. (2010)
Cement for site preparation [kg]	Site preparation	all	s	300 kg/site	Frick et al. (2010)
Diesel for drilling rig drive [MJ]	Drilling	all	s, z, n_w	7.492 MJ/m	Frick et al. (2010)
Diesel for drilling mud [MJ]	Drilling	all	s, z, n_w	181.3 MJ/m	Frick et al. (2010)
Bentonite for drilling mud [kg]	Drilling	all	s, z, n_w	7.7 kg/m	Frick et al. (2010)
Starch for drilling mud [kg]	Drilling	all	s, z, n_w	12.8 kg/m	Frick et al. (2010)
Chalk for drilling mud [kg]	Drilling	all	s, z, n_w	5.4 kg/m	Frick et al. (2010)
Water for drilling mud [kg]	Drilling	all	s, z, n_w	671.4 kg/m	Frick et al. (2010)
Calcium carbonate for drilling mud [kg]	Drilling	all	s, z, n_w	6.7 kg/m	Frick et al. (2010)
Cuttings disposal [kg]	Drilling	all	s, z, n_w	456 kg/m	Frick et al. (2010)
High-alloyed steel for casing [kg]	Casing	all	s, z, n_w	34 kg/m	Frick et al. (2010)
Low-alloyed steel for casing [kg]	Casing	all	s, z, n_w	69.1 kg/m	Frick et al. (2010)
Bentonite for cementation [kg]	Cementation	all	s, z, n_w	0.2 kg/m	Frick et al. (2010)
Portland cement for cementation [kg]	Cementation	all	s, z, n_w	23.5 kg/m	Frick et al. (2010)
Silica sand for cementation [kg]	Cementation	all	s, z, n_w	7 kg/m	Frick et al. (2010)
Unspecified cement for cementation [kg]	Cementation	all	s, z, n_w	7.3 kg/m	Frick et al. (2010)
Water (decarbonized) for cementation [kg]	Cementation	all	s, z, n_w	16.9 kg/m	Frick et al. (2010)
Diesel for reservoir enhancement [MJ]	Hydraulic stimulations	all	s, n_w	$3 \cdot 10^6$ MJ/well	Frick et al. (2010)
Water (demineralized) for reservoir enhancement [kg]	Hydraulic stimulations	all	s, n_w	$269 \cdot 10^6$ kg/well	Frick et al. (2010)
Hydrochloric acid for reservoir enhancement [kg]	Chemical stimulations	all	s, n_w	$2.225 \cdot 10^3$ $(1.5 - 3 \cdot 10^3)$ kg/well	Portier et al. (2009)
Transport by lorry associated with construction [tkm]	Construction	all	s, n_w	$144 \cdot 10^3$ tkm/ well	Frick et al. (2010)
Transport by rail associated with construction [tkm]	Construction	all	s, n_w	$413 \cdot 10^3$ tkm/ well	Frick et al. (2010)

Table 4.3 Summary of the process equipment included in the LCA model, and of the parameters used for their scaling and adaptation to the configuration (Gerber and Maréchal (2012c)).

Name of the LCI element	Life cycle stage	Conv. tech.	Functional parameters	Source
Geothermal pumps	Plant construction	all	$\dot{E}^+_{EGS}(\dot{m}_{inj},$ $dP_{EGS}),\, P_{inj}$	Gerber et al. (2011a,b)
Flash drums	Plant construction	1F, 2F	$m_f(T_{ext}(z),$ $dP_f, x_d), \dot{Q}^-_{max}$	Gerber et al. (2011a,b)
Turbines	Plant construction	all	$\dot{E}^-_{tb}(T_{ext}(z),$ $T_a, x_d), \dot{Q}^-_{max}$	Gerber et al. (2011a,b)
Cycle pumps	Plant construction	all	$\dot{E}^-_{tb}(T_{ext}(z),$ $T_a, x_d), \dot{Q}^-_{max}$	Gerber et al. (2011a,b)
Heat exchangers & condensers	Plant construction	all	$\dot{E}^-(T_{ext}(z),$ $T_a, x_d), \dot{Q}^-_{max}$	Gerber et al. (2011a,b)
District heating network	Plant construction	all	$\dot{Q}^-_{max}$	Girardin et al. (2010)
Initial amount of working fluid	Plant construction	ORCs, Kalina	$\dot{E}^-(T_{ext}(z),$ $T_a, x_d), \dot{Q}^-_{max},$ wf	Gerber et al. (2011a)

Site preparation. The quantities of auxiliary materials required for the site preparation are expressed as:

$$M_{j,sp} = \frac{v_j}{s} \tag{4.4}$$

where $M_{j,sp}$ is the required quantity of the auxiliary material j for the site preparation, v_j is its reference quantity needed per site and s is the percentage of success in achieving the EGS sub-surface plant construction in a given project. For example, $s = 50\%$ means that two sites have to be explored, drilled and enhanced before an EGS can be successfully operated. For the base case, a full success factor is assumed. The v_j values for site preparation are taken from Frick et al. (2010).

Drilling, casing and cementation. The quantities of auxiliary materials required for the drilling, casing and cementation are expressed as:

$$M_{j,dc} = \frac{v_j \cdot z \cdot n_w}{s} \tag{4.5}$$

where z is the EGS targeted construction depth, and n_w is the number of wells that have to be drilled. In the present case study, $n_w = 3$ is assumed. The v_j values for the drilling, casing and cementation are taken from Frick et al. (2010).

Reservoir enhancement. The quantities of auxiliary materials required for the reservoir enhancement are expressed as:

$$M_{j,re} = \frac{v_j \cdot n_w}{s} \tag{4.6}$$

The v_j values are taken from Frick et al. (2010) for the hydraulic stimulations, and from Portier et al. (2009) for the chemical stimulations.

Working fluid in binary cycles. For the binary power plant construction (ORCs or Kalina), the quantity of working fluid initially required is calculated from the data available in Frick et al. (2010) for iso-butane, and then adapted proportionally to the power output of the cycle and as a function of the thermodynamic properties of the working fluid.

Water for make-up. The quantity of water for make-up during EGS operation is expressed as:

$$\dot{m}_{mkup} = l_{wat} \cdot \dot{m}_{inj} \tag{4.7}$$

where l_{wat} are the water losses in the EGS, assumed to be 10% according to Minder et al. (2007). Thus, the reinjected mass flow rate $\dot{m}_{inj}$ corresponds to 100 kg/s considering the 90 kg/s assumed for the extraction well.

Scaling and residues. The quantity of scaling and residues that have to be disposed during the EGS operation is expressed as:

$$\dot{m}_{scal} = v_{scal} \cdot \dot{m}_{ext} \tag{4.8}$$

where v_{scal} is the quantity of scaling and residues per kg of geothermal water and $\dot{m}_{ext}$ is the expected mass flow rate at the extraction wells, v_{scal} being calculated using the data from Frick et al. (2010).

Working fluid losses. The yearly losses and required make-up of working fluid for binary cycles during the EGS operation are expressed as:

$$L_{wf} = M_{wf}(\dot{E}_{tb}^-(z, \dot{Q}_{max}^-, x_d), y_{wf}) \cdot l_{wf} \tag{4.9}$$

where M_{wf} is the initial required quantity of working fluid, $\dot{E}_{tb}^-$ is the electrical power produced by the turbines of the cycle, which varies with the resource temperature at depth z, the district heating network installed capacity $\dot{Q}_{max}^-$

Table 4.4 Summary of the different LCI elements included in the LCA model for operation, and of the parameters used for their scaling and adaptation to the configuration (adapted from Gerber and Maréchal (2012c)).

Name of the LCI element & Unit	Life cycle stage	Conv. tech.	Functional parameters	Reference quantity v_k	Source
Water make-up for EGS [kg]	EGS operation	all	$l_{wat}, \dot{m}_f$	from model	Gerber et al. (2011b)
Transport by lorry for scaling and residues [tkm]	EGS operation	all	–	250 tkm/yr	Frick et al. (2010)
Disposal of scaling and residues [kg]	EGS operation	all	$\dot{m}_{ext}$	1.5 kg/yr·m³/h	Frick et al. (2010)
Equipment maintenance	EGS operation	all	I_{pe}	from model	Gerber et al. (2011a)
Fossil CO_2 emitted at condenser [kg]	EGS operation	1F, 2F	$\dot{m}_f$, em_{CO_2}	from model	Baldacci et al. (2002)
H_2S emitted at condenser [kg]	EGS operation	1F, 2F	$\dot{m}_f$, em_{H_2S}	from model	Baldacci et al. (2002)
Fossil CH_4 emitted at condenser [kg]	EGS operation	1F, 2F	$\dot{m}_f$, em_{CH_4}	from model	Baldacci et al. (2002)
H_2 emitted at condenser [kg]	EGS operation	1F, 2F	$\dot{m}_f$, em_{H_2}	from model	Baldacci et al. (2002)
NH_3 emitted at condenser [kg]	EGS operation	1F, 2F	$\dot{m}_f$, em_{NH_3}	from model	Baldacci et al. (2002)
Working fluid loss for binary cycles [kg]	EGS operation	ORCs, Kalina	l_{wf}, wf, $\dot{E}_{tb,p}^-$	from model	Ormat (2010)
Working fluid make-up for binary cycles [kg]	EGS operation	ORCs, Kalina	l_{wf}, wf, $\dot{E}_{t,p}^-$	from model	Ormat (2010)
Transport for working fluid make-up [tkm]	EGS operation	ORCs, Kalina	l_{wf}, wf, $\dot{E}_{tb,p}^-, d$	30 km	distance assumed
Avoided electricity from NGCC [kWh$_e$]	EGS operation	all	$\dot{E}_p^-(z, x_d)$, $\dot{Q}_p^-$	from model	Gerber et al. (2011b)
Avoided district heating from natural gas condensing boiler [MJ$_{th}$]	EGS operation	all	$\dot{E}_p^-(z, x_d)$, $\dot{Q}_p^-$	from model	Gerber et al. (2011b)

and the design decision variables of the conversion system x_d. y_{wf} are the thermodynamic properties of the chosen working fluid and l_{wf} are the yearly losses of the working fluid, which are between 0 and 2% according to Ormat (2010). The maximum value of 2% is thus assumed in the present study.

Emissions from flash systems. The emissions of non-condensable gases from the condenser for flash systems during the EGS operation are expressed as:

$$\dot{m}_{i,em} = em_i \cdot \dot{m}_f(\dot{E}_{tb,p}^-(z, \dot{Q}_p^-, x_d)) \qquad (4.10)$$

where em_i is the emission factor of the non-condensable gas i, in kg of substance i per kg of geothermal water $\dot{m}_f$ passing through the flash system turbine and the condenser, function of the electricity produced by the turbines at period p, which is itself function of the depth and of the district heating required at period p, $\dot{Q}_p^-$.

No data are currently available for potential emissions from flash systems using EGS. Thus, average data for hydrothermal systems from Baldacci et al. (2002) are used considering the emissions of CO_2, H_2S, CH_4, H_2 and NH_3. For CO_2, emissions are as well available per MWh_e of produced electricity from other sources (DiPippo (2008); Brown and Ulgiati (2002); Frondini et al. (2009)), which allows for estimating the emissions per kg of geothermal steam and comparing them with the ones given in Baldacci et al. (2002). While the estimations calculated from Frondini et al. (2009) (0.045-0.081 kg-CO_2/kg-steam) are in the same range than the values in Baldacci et al. (2002) (0.037 kg-CO_2/kg-steam), the estimation calculated from the data of DiPippo (2008) (0.005 kg-CO_2/kg-steam) is one order of magnitude below, and is thus assumed as the minimum value, while the estimation from the data of Brown and Ulgiati (2002) (0.128 kg-CO_2/kg-steam) is one order of magnitude above, and is taken as the maximum value.

Due to the different geochemistry of EGS and hydrothermal systems, these values should however be updated once data are available for emissions from flash systems combined with EGS.

Table 4.5 Summary of the different LCI
elements included in the LCA model for end-of-life, and of the parameters used for their scaling and adaptation to the configuration (adapted from Gerber and Maréchal (2012c)).

Name of the LCI element & Unit	Life cycle stage	Conv. tech.	Functional parameters	Reference quantity v_k	Source
Gravel for well dismantling [kg]	Dismantling	all	s, z, n_w	51.1 kg/m	Frick et al. (2010)
Cement for well dismantling [kg]	Dismantling	all	s, z, n_w	4.9 kg/m	Frick et al. (2010)
Equipment disposal	Dismantling	all	r	from model	Gerber et al. (2011a)
Working fluid loss at dismantling [kg]	Dismantling	all	$l_{wf,e}$	from model	Saner et al. (2010)

End-of-life. For the end-of-life phase, the v_j quantities for cement and gravel used for well decommissioning are taken from Frick et al. (2010), and are as well expressed as a function of the depth of the well by using Equation (4.5). No data are available for the working fluid losses from the binary cycle during the dismantling of the power plants. Therefore, the value for losses from heat pump systems corresponding to 20%, which is reported in Saner et al. (2010), is used.

Table 4.6 Description of the parameters used to calculate the LCI elements (adapted from Gerber and Maréchal (2012c)).

Abbreviation	Description	Unit	Range or quantity (if fixed)
s	Percentage of success in EGS construction	–	1
z	Construction depth of EGS	m	3000-10 000
n_w	Number of wells	–	3
$\dot{E}^+_{EGS}$	Electricity consumed by the geothermal pumps	kW	from model
$\dot{m}_{inj}$	Injected water mass flow rate in EGS	kg/s	100
dP_{inj}	Pressure drop inside EGS	bar	100
P_{inj}	Pressure at which water is injected[a]	bar	from model
m_f	steam mass flow rate passing through the flash turbines	kg/s	from model
T_{ext}	Temperature at extraction well	°C	from model
dP_f	Pressure drop inside flash drums	bar	from model
$\dot{Q}^-$	District heating requirements	kW	0-60 000
$\dot{E}^-_{tb}$	Electricity produced at turbine	kW	from model
$\dot{E}^-$	Net electricity produced by the conversion system[b]	kW	from model
T_a	Temperature of the ambiance	°C	10
wf	Choice of working fluid	–	see Table 4.1
l_{wat}	Percentage of water losses inside EGS	–	0.1
$\dot{m}_{ext}$	Water mass flow rate at extraction well	kg/s	90
I_{pe}	Yearly percentage of initial impact for process equipment	–	0.05 (Gerber et al. (2011a))
em_{CO_2}	Specific CO_2 emissions per unit of flashed geothermal steam	kg-CO_2/kg-steam	0.0366 (Baldacci et al. (2002))
em_{H_2S}	Specific H_2S emissions per unit of flashed geothermal steam	kg-H_2S/kg-steam	$4.85 \cdot 10^{-4}$ (Baldacci et al. (2002))
em_{CH_4}	Specific CH_4 emissions per unit of flashed geothermal steam	kg-CH_4/kg-steam	$2.6 \cdot 10^{-4}$ (Baldacci et al. (2002))
em_{H_2}	Specific H_2 emissions per unit of flashed geothermal steam	kg-H_2/kg-steam	$2.75 \cdot 10^{-5}$ (Baldacci et al. (2002))
em_{NH_3}	Specific NH_3 emissions per unit of flashed geothermal steam	kg-NH_3/kg-steam	$1.175 \cdot 10^{-4}$ (Baldacci et al. (2002))
l_{wf}	Yearly percentage of initial amount of working fluid lost in the atmosphere	–	0.02 (Ormat (2010))
d	Transport distance for working fluid	km	50
$l_{wf,e}$	Percentage of working fluid lost during plant dismantling	–	0.2 (Saner et al. (2010))
r	Recycling ratio for each type of process equipment	–	0.5-0.98 (Gerber et al. (2011a))

[a] calculated to keep water in liquid state at the temperature corresponding to depth z after pressure drop.

[b] after removal of parasitic losses (geothermal pumps, cycle pumps).

Life Cycle Impact Assessment

Here, two different impact assessment methods are used: the method of the Intergovernmental Panel on Climate Change (2007) (IPCC07), which is used to quantify the global warming potential on a 100-year time-horizon in terms of CO_2-equivalents, and the Ecoindicator99-(h,a) (Goedkoop and Spriensma (2000)). To calculate the total impact over the life cycle of the EGS, Equation (2.23) is used, with the denominator being equal to one, since the functional unit is one EGS accounting for its full life cycle.

4.3 Environomic optimal configurations

The goal of the multi-objective optimization (MOO) is to identify the optimal configurations of geothermal conversion systems for EGS considering different potential combinations of technologies, resource depths and district heating network installed capacities. Economic, thermodynamic and environmental criteria have to be considered.

4.3.1 Multi-objective optimization problem

For the first step, three independent optimization objectives are selected:

1) The investment costs, to be minimized:

$$Min\ CI_{tot} = \frac{CI_{EGS}(z)}{s} + \sum_{v=1}^{n_v} max(CI_v(z, \dot{Q}^-_{max}, x_d))_p + CI_{DH}(\dot{Q}^-_{max})$$

(4.11)

where CI_{EGS} are the investment costs linked with the sub-surface plant construction, varying with the targeted exploitation depth z, accounting for the success factor s. CI_v is the investment cost of the equipment v, calculated for each period p and for which the maximal value is taken, varying with z, with the design size of the district heating $\dot{Q}^-_{max}$ and with the other decision variables of the non-linear master problem x_d. CI_{DH} is the investment cost of the district heating network expressed as a function of its nominal installed capacity.

2) The annual revenue, to be maximized:

$$Max\ R_{an} = \sum_{p=1}^{n_p} t_p \cdot (c_e^- \cdot \dot{E}_p^-(z, \dot{Q}^-_{max}, x_d) + c_{DH}^- \cdot \dot{Q}_p^-(\dot{Q}^-_{max})$$
$$-\dot{co}_{EGS}(z) - \sum_{t=1}^{n_t} \dot{co}_t(z, \dot{Q}^-_{max}, x_d))$$

(4.12)

where t_p is the operating time associated with period p, c_e^- and c_{DH}^- are the specific selling prices of electricity and district heating, respectively, $\dot{E}_p^-$ is the net electrical power produced at the operating conditions of period p –

the consumption of geothermal pumps and cycle pumps is accounted for – $\dot{Q}_p^-$ is the district heating power supplied at period p, calculated considering the installed capacity of the district heating network and the corresponding demand for the period, $\dot{co}_{EGS}$ is the specific operating cost of the EGS and $\dot{co}_t$ is the specific operating cost of conversion technology t. c_e^- and c_{DH}^- are assumed to be 0.16 \$/kWh$_e$ and 0.11 \$/kW$_{th}$, respectively, which is representative of the average Swiss market conditions (OFEN (2010b)), assuming no variation in the electricity and district heating prices over the EGS expected lifetime.

3) The exergy efficiency of the conversion system, representing the ratio between the exergy services supplied and the exergy from the EGS entering the conversion system, to be maximized:

$$Max\ \eta = \frac{\sum_{p=1}^{n_p} t_p \cdot (\dot{E}_p\,(z, \dot{Q}_{max}^-, x_d) + \dot{Q}_p^-(\dot{Q}_{max}^-) \cdot (1 - \frac{T_a}{T_{DH,lm,p}}))}{\sum_{p=1}^{n_p} t_p \cdot \dot{Q}_{EGS,p}^+(z, x_d) \cdot (1 - \frac{T_a}{T_{EGS,lm,p}})} \tag{4.13}$$

where $\dot{Q}_{EGS,p}^+$ is the available thermal power from the EGS at period p, T_a is the ambient temperature, assumed to be $10°C$, and T_{lm} is the logarithmic mean temperature of the heating or cooling required in $°K$, calculated by:

$$T_{lm} = \frac{T_{in} - T_{out}}{ln(\frac{T_{in}}{T_{out}})} \tag{4.14}$$

where T_{in} is the inlet temperature of the hot source and T_{out} is the outlet temperature at which the hot source is cooled down. For the district heating, T_{in} is the supply temperature and T_{out} is the return temperature. For the EGS, T_{in} is the temperature at extraction well and T_{out} is the reinjection temperature.

For each possible combination of conversion technologies and each proposed working fluid in the binary cycles, the trade-off between the three objectives is calculated by generating a Pareto front using the genetic algorithm (Molyneaux et al. (2010)). The decision variables considered for the optimization problem are given in Table 4.7.

Except for the depth z and the district heating installed capacity $\dot{Q}_{max}^-$, all the decision variables are generated using the multi-period strategy, meaning that each variable is allowed to take a different value for each operation period. Indeed, even if the size of the district heating network is fixed, it corresponds to a seasonal variation in the requirements. Using a multi-period strategy to generate the decision variables allows for adapting the system operation to the seasonal district heating requirements. The advantages of using such an approach are discussed in details in Gerber and Maréchal (2012a).

Table 4.7 Decision variables used for the MOO of the different combinations of conversion technologies.

Name	Conversion technology	Range	Unit
EGS construction depth z	all	[3000; 10 000]	m
District heating network installed capacity $\dot{Q}_{max}^-$	all	[0; 60]	MW_{th}
Reinjection temperature of geothermal water	all	[70; 130]	°C
Expansion ratio in first flash drum	1F, 2F	[0; 1]	–
Expansion ratio in second flash drum	2F	[0; 1]	–
Evaporation temperature of working fluid	ORC, ORC-d, ORC-2	[60; 135][a]	°C
Temperature difference between geofluid and superheated working fluid	ORC, ORC-d, ORC-2, ORC-s, Kalina	[5; 20]	°C
Temperature at which intermediate draw-off is done	ORC-d	[20; 65]	°C
Splitting fraction between district heating and additional electricity production	ORC-d	[0;1]	–
Evaporation temperature of working fluid at lower pressure	ORC-2	[60; 135][a]	°C
Splitting fraction between higher and lower pressure evaporation	ORC-2	[0;1]	–
Higher pressure of the supercritical working fluid	ORC-s	[34;80][a]	bar
Evaporation pressure of the working fluid	Kalina	[32;42]	bar
Condensation pressure of the working fluid	Kalina	[6;10]	bar
Ammonia molar fraction in the working fluid	Kalina	[0.7;0.85]	–

[a] adapted to the chosen working fluid.

4.3.2 Selection of final optimal configurations

Since the multi-objective optimization results in a large number of optimal points, each one representing one possible configuration for the geothermal system, a selection of configurations has to be performed to classify the solutions. This is done for each cluster of technologies and working fluid by selecting one configuration each 500m between 3000 and 10 000m, and each 5 MW_{th} for the district heating network capacity from 0 to 60 MW_{th} – 0 MW_{th} meaning single electricity production. To select the final configuration at a given depth and district heating capacity, the payback period of the overall system is used, and other associated thermodynamic and environmental indicators are as well calculated.

The simple payback period is calculated by:

$$t_{pb} = \frac{CI_{an}}{R_{an}} \tag{4.15}$$

where R_{an} is the annual revenue calculated with Equation (4.12), considering a constant price for the energy services, and CI_{an} are the annualized investment costs:

$$CI_{an} = CI_{tot} \cdot \frac{ir \cdot (ir + 1)^{t_{yr}}}{(ir + 1)^{t_{yr}} - 1} \tag{4.16}$$

where CI_{tot} are the total investment costs calculated with Equation (4.11), ir is the interest rate, assumed to be 6%, and t_{yr} is the lifetime of the EGS, assumed to be 30 years.

For analyzing the thermodynamic performance, the yearly exergy efficiency of the conversion system, which has been detailed previously in Equation (4.13), is used. It has to be pointed out that this formulation for the exergy efficiency accounts for the exergy delivered by the geothermal heat source with a varying reinjection temperature.

For the environmental performance, two indicators are used:

1) the yearly avoided CO_2-equivalent emissions, using the IPCC07 impact assessment method. The life cycle CO_2 emissions for construction, operation and end-of-life are compared with the production of the same services with a NGCC for electricity and a natural gas boiler for heating:

$$I_{CO_2,av} = \sum_{p=1}^{n_p} (t_p \cdot (\dot{E}_p^- \cdot I_{CO_2,NGCC} + \dot{Q}_p^- \cdot I_{CO_2,NGB}$$

$$- \sum_{j=1}^{n_{j,o}} \dot{IO}_{j,p})) - \frac{\sum_{j=1}^{n_{j,c}} max(IC_j)_p + \sum_{j=1}^{n_{j,e}} max(IE_j)_p}{t_{yr} \cdot s} \tag{4.17}$$

where $I_{CO_2,NGCC}$ and $I_{CO_2,NGB}$ are the specific CO_2 equivalent emissions of electricity production from NGCC and for heating production from a condensing natural gas boiler, respectively, taken from ecoinvent®, and are equal to 0.425 kg CO_2-eq/kWh$_e$ and 0.241 kg CO_2-eq/kWh$_{th}$, respectively.

2) the relative life cycle avoided impacts, using the single-score of the Ecoindicator99-(h,a). The impacts are again compared with the production of the same services. The best configuration of all is fixed as the reference (100% of avoided impacts), and the other ones are then compared with this value.

4.3.3 Results and discussion

Pareto curves

Figure 4.12 shows some of the Pareto curves obtained for the tri-objective thermo-economic optimization. Accounting for the different configurations and working fluids, 38 Pareto curves were generated. Therefore, only a fraction of the potential combinations of technologies are displayed as examples for readability, to illustrate the behavior of the system configurations in the optimization: one with a single-flash system, one with a binary cycle – an ORC with an intermediate draw-off – and one with a single-flash system and a bottoming ORC.

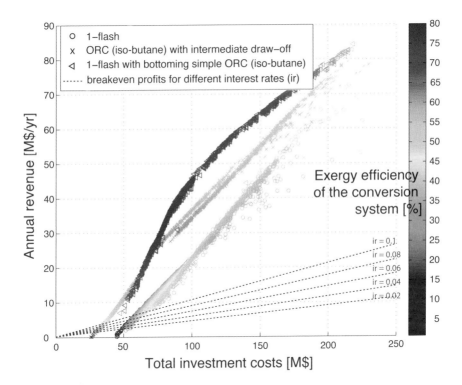

Figure 4.12 Examples of Pareto curves obtained by the tri-objective thermo-economic optimization, with the corresponding break-even annual revenue for different interest rates (adapted from Gerber and Maréchal (2012c)).

All the curves show a net trade-off between the investment costs and the annual revenue, and, in most of the cases, another trade-off between the exergy efficiency of the conversion system and the economic objectives. The effects of the EGS depth and of the district heating size on the trade-off between the investment and the revenue, and on the other performance indicators are discussed in the next subsection.

Final optimal configurations

The final optimal configurations are then selected from these Pareto curves, on the basis of the minimal payback period, detailed in Equation (4.15), for a varying EGS construction depth and district heating network installed capacity. The results are displayed in Figure 4.13. The available thermal power from the EGS as a function of the depth, assuming a minimal reinjection temperature of $70°C$, is as well displayed. The associated exergy efficiencies, the avoided CO_2 emissions and the relative avoided life cycle impacts with Ecoindicator99-(h,a) are displayed in Figures 4.17, 4.18, and 4.20, respectively. For a more detailed analysis, supplementary results are available in Annex C.

The environomic performances of the optimal configurations are analyzed in details in the following paragraphs, with an emphasis on the life cycle environmental indicators. The similarities or the trade-offs between the different considered criteria are as well highlighted.

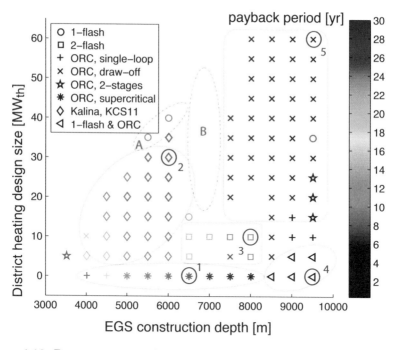

Figure 4.13 Best conversion technologies selected on the basis of the payback period, as a function of EGS depth and design size of the district heating (adapted from Gerber and Maréchal (2012c)).

From Figure 4.13, it appears first that with the economic assumptions and the geological conditions taken for this application case study, deeper EGS from 7000m to 10 000m are economically more attractive. However, some of the Kalina configurations for shallower wells also have a low payback period, due to the system performance in cogeneration mode. From 7500m, the payback period does not decrease significantly anymore, despite the increased production of energy services. This is explained by the non-linear increase of the drilling costs of the EGS (see Figure 4.3). In the case of a shallow EGS down to 6000m, the increase in the district heating installed capacity decreases the payback period and makes therefore CHP more competitive than single electricity production. In the case of deep EGS from 7500m to 10 000m, there is no important change in the payback period with the increase of the district heating installed capacity, though increased district heating requirements decrease the electricity production. Regarding deep EGS, economics is therefore not necessarily the most important criterion for the decision-making. In consequence, thermodynamic and environmental criteria can influence a lot on the final decision.

A detailed cost-benefit analysis of five typical configurations is displayed in Figure 4.14, for the configurations identified by a red circle on Figure 4.13. The detailed operating conditions and the associated integrated exergy composite curves (Maréchal and Kalitventzeff (1996)) of these five configurations are available in Annex C.

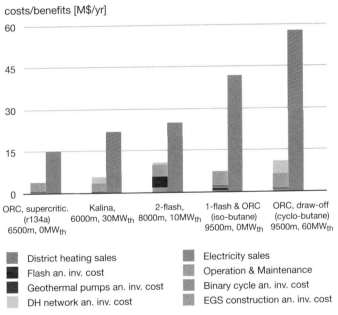

Figure 4.14 Cost-benefit analysis on a yearly basis of 5 typical configurations from Figure 4.13 (adapted from Gerber and Maréchal (2012c)).

On Figure 4.13, five major zones can be distinguished, each one of them illustrated by a configuration in Figure 4.14:

1) EGS (4000-8000m) for electricity production only with an ORC. The zone corresponding to these configurations is characterized by the lowest investment costs, but also by the lowest profits. The selected best technology is

an ORC with a single-loop using R134a in the shallowest depths, and at larger depths a supercritical ORC still using R134a, except for one configuration using n-pentane, as it can be seen in Figure 4.15 showing the working fluid selection for the ORC configurations.

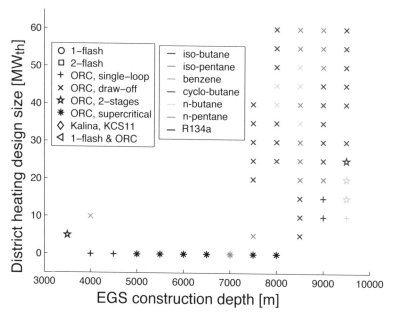

Figure 4.15 Working fluid selection associated with the best configurations of Figure 4.13, for the configurations using an ORC.

2) EGS (4500-6000m) for cogeneration (5-35 MW$_{th}$) with a Kalina cycle. In the case of these configurations, the payback period decreases with an increased installed capacity of the district heating network, since it plays the role of the cold source in the Kalina cycle producing electricity, as illustrated by the exemplary integrated exergy composite curve of the Kalina cycle in Figure 4.16. This curve allows for representing the heat exchanges between the cycle and the other components of the system – EGS, district heating and cold source – in terms of exergy, using the Carnot factor on the y-axis. There is therefore no trade-off between electricity and heat in the case of the Kalina cycles. Two configurations in this zone with a large district heating network (zone A on Figure 4.13) use a flash system. Indeed, at these depths, the temperature is rather low for electricity production using a flash system. Thus, using the waste heat from the liquid part of the flash drum for district heating before reinjection allows for a significant increase in the efficiency and makes this technology competitive.

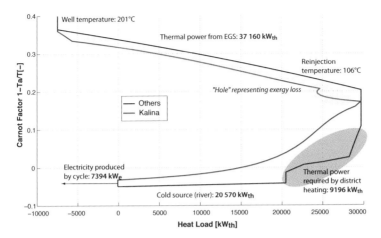

Figure 4.16 Integrated exergy composite curve of the configuration using a Kalina cycle with a district heating design size of 30 MW at 6000m operating during interseason (adapted from Gerber and Maréchal (2012c)).

3) EGS (6500-8000m) for cogeneration (5-15 MW_{th}) with a flash system. In this range of depths, large CHP systems (zone B on Figure 4.13) are not interesting when compared with electricity production or small CHP systems. The Kalina cycle is not selected anymore by the optimizer, since the exergy losses increase too much beyond 6000m – the yellow area in the integrated composite curve of the cycle, in Figure 4.16. For other cycles, including flash systems for which the temperatures become thermodynamically favorable to electricity production, the configurations at these depths were suboptimal and thus not kept by the optimizer. Indeed, for a configuration in this range of depths with a large district heating network, better exergy efficiency and revenues can be achieved for the same investment costs by drilling deeper to reach a higher temperature and favoring electricity production, which saves the investment of the district heating network. The temperature here is sufficiently high for using flash systems, which are favored for small cogeneration systems since it uses part of the waste heat before reinjection without penalizing the electricity production. The investment linked with a large flash system is however important.

4) EGS (8500-10 000m) for electricity production only with a flash system and a bottoming ORC. In this range, the heat from the liquid part of the flash system is sufficient to use a bottoming ORC, which allows for increasing the electricity output and decreasing as well the size of the flash system and its related investment.

5) EGS (7500-10 000m) for cogeneration (10-60 MW_{th}) with an ORC. These are the configurations with the largest investment costs due to the depth of the EGS and to the district heating capacity, but also the ones generating the highest profits, which is why they were kept by the optimizer, even if they do not increase the exergy efficiency and increase the investment

costs. The selected technology is in majority an ORC with an intermediate draw-off, using the district heating as part of the cold source. Different working fluids are selected depending on the depth and on the district heating requirements, as it can be seen in Figure 4.15.

The exergy efficiency of the conversion system associated with the optimal economic configurations is displayed in Figure 4.17.

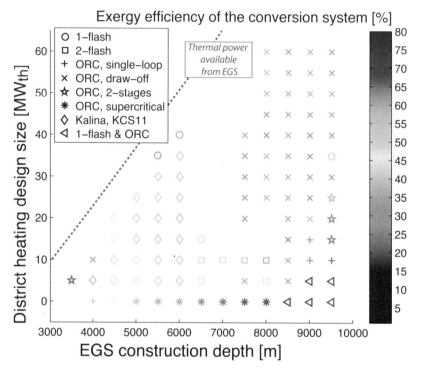

Figure 4.17 Exergy efficiencies of the conversion system associated with the best configurations of Figure 4.13 (adapted from Gerber and Maréchal (2012c)).

The exergy efficiency of the conversion system depends on the depth and on the district heating capacity. The highest efficiencies of about 75% are achieved with a deep EGS between 8500m and 9000m using a single-flash system with a bottoming ORC, almost exclusively for electricity production. In the case of CHP systems, the highest efficiencies of about 60% are achieved by an ORC with an intermediate draw-off between 7500m and 9500m. For CHP, the exergy efficiency reaches a maximum around 7500m and remains then relatively constant, due to the switches in the choice of the working fluid as a function of the depth, as it can be seen in Figure 4.15.

The yearly-avoided CO_2 emissions associated with the optimal economic configurations are displayed in Figure 4.18. A detailed CO_2 balance is displayed in Figure 4.19 for the configurations identified by a red circle on Figure 4.18.

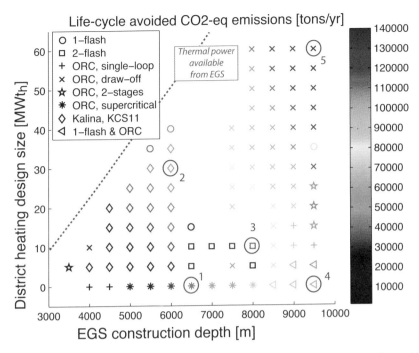

Figure 4.18 Yearly avoided CO_2 emissions associated with the best configurations of Figure 4.13, (adapted from Gerber and Maréchal (2012c)).

The yearly-avoided CO_2 emissions, calculated on a life cycle basis, increase with the EGS depth. Though there is a high variation between the shallowest and the deepest configuration, none of the selected optimal configurations has a negative CO_2 balance. Like for the economic calculations, this is however only valid for the geological conditions assumed in the present case. As shown by Figure 4.19, when compared with the beneficial impacts from the substitution of energy services, the harmful impacts due to the construction of the EGS and of the power plant are insignificant in the present case study.

From 4000m to 6000m and from 7500m and 9500m, the avoided CO_2 emissions associated with a certain depth increase with the district heating capacity. Examples are the configuration 2 with a Kalina cycle and the configuration 5 with an ORC integrating an intermediate draw-off, which has the highest avoided CO_2 emissions. On the contrary, between 6000m and 7500m, the configurations for single electricity production with a binary cycle, like the configuration 1 with a supercritical ORC, have higher avoided CO_2 emissions than CHP systems using a flash system, like the configuration 3, since flash systems directly use the geothermal steam that contains CO_2 and other gases. This geothermal steam is then released to the atmosphere at the condensers with the non-condensable gases. As shown by the maximum potential CO_2 emissions on the graph, the balance might even be negative if EGS using flash systems have CO_2 emissions that are comparable to the maximum value of some existing hydrothermal systems. However, since the data used here to

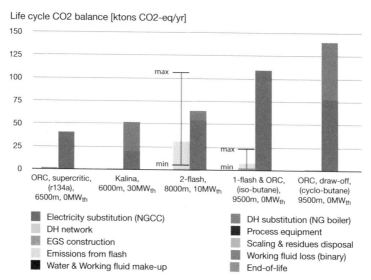

Figure 4.19 CO_2-equivalent balance on a yearly basis for 5 typical configurations of Figure 4.13, (adapted from Gerber and Maréchal (2012c)). Minimum and maximum values for the CO_2 emissions from the flash condensers are shown.

calculate these emissions are not for EGS, this particular aspect has to be verified once reliable data are available for the emissions from such systems. In the present case, the use of a bottoming binary cycle with a single-flash system, as illustrated by configuration 4, allows for increasing significantly the electricity output, and for decreasing the emissions from the flash, which has a smaller size and uses therefore less steam.

Instead of using the linear model of Frick et al. (2010) for the impact scaling as a function of the depth for the EGS construction and its end-of-life, the exponent used for the non-linear scaling of the drilling costs, equal to 1.3879, can be used. Indeed, the analogy between the costs and the environmental impacts has been validated for process equipment earlier (see subsection 2.2.4) and could as well probably be applied for the auxiliary materials depending on the EGS depth. However, the difference in the total avoided CO_2 emissions with the linear scaling is at the most an increase of 1.5% if the power scaling is applied, the impact being dominated by the substitution of energy services and the emissions from the flash systems.

The relative life-cycle avoided impacts calculated with Ecoindicator99-(h,a), in percentage of the best configuration (9500m and 60MW$_{th}$), are presented in Figure 4.20. A detailed balance for life cycle impacts is displayed in Figure 4.21 for the configurations identified by a red circle in Figure 4.20.

Here again, though the relative differences are important, no configuration has a negative environmental balance. The avoided impacts show a similar behavior to the avoided CO_2 emissions, increasing with depth and with district heating

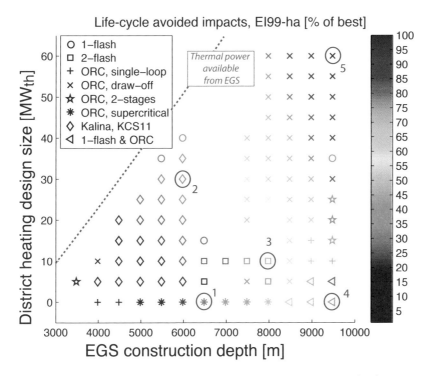

Figure 4.20 Relative life cycle avoided impacts with Ecoindicator99-(h,a) associated with the best configurations of Figure 4.13, (adapted from Gerber and Maréchal (2012c)).

capacity. However, unlike for the avoided CO_2 emissions, there is a less clear difference between flash systems and binary cycles in favor of the latter, because the Ecoindicator99-(h.a) weighs more strongly the substitution of natural gas. Therefore, the impact of CO_2 emissions from flash systems are diluted, as it can be seen in Figure 4.21.

These two environmental criteria can be compared with the other economic and thermodynamic performance indicators, the payback period and exergy efficiency of the conversion system, respectively. In the shallow range of EGS depths, from 4000m to 6000m, all the decision criteria favor the EGS at 6000m with a Kalina cycle and a large district heating network with an installed capacity between 20 and 35 MW_{th}.

However, in the deepest range of EGS depths, between 7500m and 9500m, integrating environmental criteria in the decision-making procedure leads to favor solutions that would not be considered if only the payback period and the exergy efficiency of the conversion system were considered. Indeed, the exergy efficiency and the payback period lead rather to favor the single electricity production over CHP systems, while environmental criteria both lead to favor CHP systems with large district heating networks. Though electricity produc-

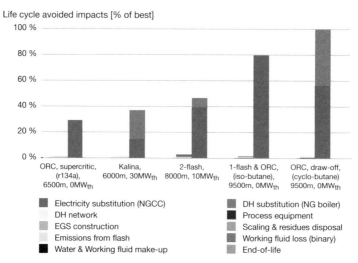

Figure 4.21 Life-cycle avoided impacts with Ecoindicator99-(h,a), in percentage of best configuration, for 5 typical configurations of Figure 4.13.

tion avoids more CO_2 than district heating on the basis of the kWh delivered, CHP systems with large district heating networks have higher energy efficiencies due to an increased heat production, and therefore avoid more CO_2 than the single electricity production. The trade-off between electricity production and district heating can be seen by comparing Figures 4.22, 4.23 and 4.24, showing the energy efficiencies, the electrical efficiencies and the efficiencies for district heating production for the EGS optimal configurations, respectively. This prevalence of large CHP systems over single electricity production for deep EGS is as well true for the impact calculated with Ecoindicator99-(h,a).

If the best configuration from an environmental point of view, the configuration 5 (ORC-d at 9500m and 60 MW_{th}) is compared in relative terms with the configuration 4 (flash system and bottoming ORC at 9500m for single electricity production) the penalty is of 11% for the payback period and of 17% in terms of exergy efficiency. However, the relative improvements in terms of environmental performance are more important: 37% for the avoided CO_2 emissions, and 31% for the avoided life-cycle impacts calculated with Ecoindicator99-(h,a). Therefore, one could argue that the environmental performance outweigh the economic and thermodynamic ones, and that the configurations with CHP using a large district heating network capacity should be favored for the construction of deep EGS.

Coverage in energy services

Another way of presenting these results is to define an area where given quantities of heat and electricity have to be produced, and to calculate the resulting CO_2 emissions of these different EGS configurations, assuming that for the remaining quantities not produced by the EGS, the NGCC and the natural gas

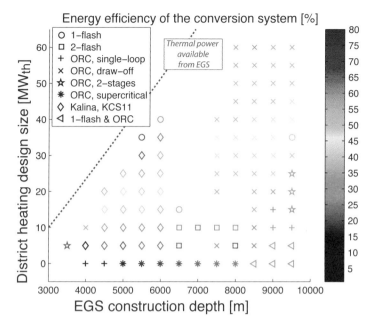

Figure 4.22 Yearly energy efficiency (including electricity and district heating) associated with the best configurations of Figure 4.13 (adapted from Gerber and Maréchal (2012c)).

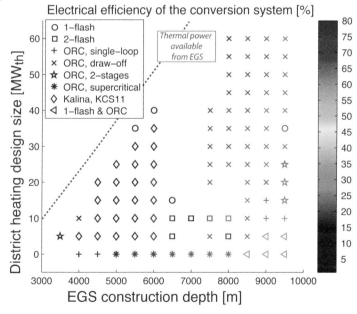

Figure 4.23 Yearly electrical efficiency (net electricity production only) associated with the best configurations of Figure 4.13 (adapted from Gerber and Maréchal (2012c)).

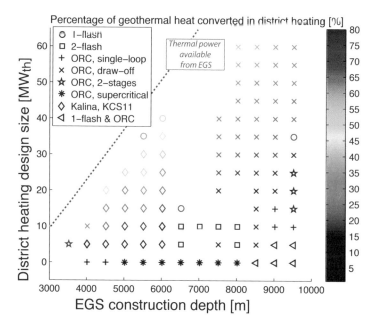

Figure 4.24 Yearly efficiencies for the conversion of geothermal heat in district heating associated with the best configurations of Figure 4.13 (adapted from Gerber and Maréchal (2012c)).

boiler are used. The maximum district heating capacity of 60 MW$_{th}$ corresponds roughly to an area of about 24 000 inhabitants. The resulting average electricity required, calculated from the data available in OFEN (2010c), corresponds to 0.875 kW$_e$ per capita, considering the total electricity consumption of the country.

Therefore, it is possible to calculate the coverage in the area for the five EGS configurations from Figure 4.13, which are presented in Table 4.8, as well as the resulting yearly CO$_2$ emissions per capita, which are presented in Figure 4.25. The scenario with separate production of heat and electricity, using a natural gas boiler and a NGCC, is as well represented in Figure 4.25, in order to assess the reduction in CO$_2$ emissions by constructing an EGS.

Table 4.8 Coverage by EGS in heat and electricity supply for the five typical configurations in an area of 24 000 inhabitants.

Configuration	Electricity coverage	Heat coverage
ORC-s (r134a), 6500m, 0MW$_{th}$	52%	0%
Kalina, 6000m, 0MW$_{th}$	26%	50%
2-flash, 8000m, 10MW$_{th}$	69%	17%
1-flash & ORC, 9500m, 0MW$_{th}$	140%	0%
ORC-d, 9500m, 60MW$_{th}$	98%	100%

At relatively shallow depths – configurations 1 and 2 – it is already possible to cover at least 50% of the requirements either in electricity or in district heating in the area by constructing an EGS. In the case of deep EGS for single electricity production, more than the requirements in electricity of the considered area are produced from geothermal heat. In terms of coverage for both services, the best configuration remains the deep EGS with the district heating installed capacity of 60 MW$_{th}$, which covers as well almost all the requirements in electricity of the area, in addition to the district heating.

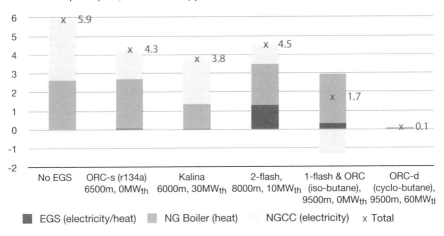

CO2 emissions per capita [tons of CO2-eq/yr]

Figure 4.25 Yearly CO_2-equivalent emissions per capita for separate production and the five typical configurations, considering the production of heat and electricity for an area of 24 000 inhabitants.

The reference scenario, not using an EGS but only the natural gas boiler and the NGCC, results in 5.9 tons of CO_2 per capita and per year, which is quite large when compared with the Swiss average of 7 tons per capita, this last value accounting for all the services to be supplied, including transport. Though in all the cases the construction of the EGS reduces in an important way the CO_2 emissions, with the minimum benefit being a reduction of 1.4 ton/cap/yr, the last configuration of the EGS at 9500m with an ORC and a district heating network with an installed capacity of 60 MW$_{th}$ is the only one which results in almost insignificant CO_2 emissions – 0.1 ton/cap/yr – for the production of electricity and heating.

It should be emphasized that these results are valid only when EGS and natural gas are the considered resources to supply heating and electricity services, without integrating any alternative that would use another source of renewable energy.

4.4 Conclusions on geothermal systems

This chapter has presented the application of the developed methodology for the integration of LCA in the design of renewable energy systems to identify the optimal environomic configurations of a mature EGS technology for Combined Heat and Power (CHP) production in the geological and economic context of Switzerland. The results of the case study in terms of economic, thermodynamic and environmental performances reflect the possible variety of the system designs in terms of EGS targeted construction depth, installed capacity of the district heating network, choice of the conversion cycle and operating conditions.

Though the final optimal configurations were selected on the basis of an economic criterion, they all have a beneficial environmental balance, both in terms of avoided CO_2-equivalent emissions and life-cycle avoided impacts, due to the output of energy services, which more than compensates the harmful impacts from the EGS construction and operation. This high efficiency in the production of energy services is due to the adopted multi-period approach, which allows for adapting the operation of CHP systems to the seasonal demand in district heating. However, the variations in the performances among the optimal configurations are important, depending on the EGS construction depth, on the district heating size and on the technology choice.

In the shallowest range of depths from 3500m down to 6000m, which could represent the first generation of a commercially mature EGS technology, the economic, thermodynamic and environmental performances all favor the configurations with a Kalina cycle, an EGS construction depth around 5500m and 6000m, and a district heating network capacity between 20 and 35 MW_{th}. Compared to the solutions with CHP, the single electricity production has less attractive performances.

In deeper ranges of depth between 7500m and 9500m, which could represent the second generation of a commercially mature EGS technology, the payback period of the system decreases slightly for systems with single electricity production using a flash system with a bottoming ORC. However, the variations in terms of exergy efficiency and environmental impacts are more important. If the environmental criterion is considered, the deepest EGS with the largest district heating network should be favored, using an ORC with an intermediate draw-off. Though the payback period of this configuration is 11% higher and the exergy efficiency 17% lower than the configuration at the same depth for single electricity production, the environmental benefits are higher: 37% of increase in the avoided CO_2 emissions and 31% for the avoided impacts considering Ecoindicator99-(h,a). In a decision-making procedure, this higher environmental benefit could play in favor of CHP configurations with a large district heating network.

Therefore, the case study demonstrates that integrating the LCA in the decision-making procedure can lead to take different decisions and consider configurations that would not be kept otherwise, when compared with the conventional thermo-economic analysis. Furthermore, it is shown that although a proper system design and integration maximizing the energy services output are crucial for enhancing the environmental performance, the thermodynamic and environmental optimums are not necessarily corresponding. Though here this is the case for shallow EGS, for deep EGS the exergy efficiency and the two environmental performance indicators are conflicting.

Finally, the simple analysis for the coverage in electricity and district heating by the EGS for a given area arises the question of the adequacy between the production of energy services from a given resource and the actual demand for these services. However, a detailed study of this problematic has to account for the other potential competing resources and technologies. Hence, this is beyond the application of the developed methodology to the design of geothermal energy conversion systems only, and this broader issue is specifically addressed in the next chapter.

CHAPTER 5

Application to urban systems

5.1 Industrial ecology applied to urban systems

Industrial ecology aims at identifying, in a given system, the potential energy and material exchanges that allow for mitigating the use of resources and the environmental impacts of human activities, in order to design sustainable economies (Allenby and Richards (1994); Ehrenfeld (1997); Erkman (1997)). The closing of material loops by the exchange of waste, by-products and energy among different industries is termed as an industrial symbiosis, by analogy with a natural ecosystem (Ehrenfeld and Gertler (1997); Chertow (2000)). The concept of industrial symbiosis can be applied to industrial complexes but also to urban systems or territories (Ehrenfeld and Gertler (1997); Singh et al. (2007); Berkel et al. (2009)).

When analyzing or designing an industrial symbiosis, both economic and environmental criteria have to be considered simultaneously (Chertow and Lombardi (2005); Jacobsen (2006)). Environmental aspects can be accounted for by integrating the LCA or related tools in the design of such systems (Singh et al. (2007); Mattila et al. (2010, 2012)). Specifically regarding the design of industrial symbioses, several methodologies have already been developed for the decision-making and the optimization of material and energy flow exchanges (Urban et al. (2010); Cimren et al. (2012)), but not for a systematic design of urban energy conversion systems.

Therefore, designing an urban energy system as an industrial symbiosis is an interesting application of the methodology presented in section 2.3 for the systematic identification of industrial ecology possibilities and supply chain synthesis. The design of such a system has to integrate the potential indigenous renewable resources, the multiple energy services to be supplied and the waste to be treated, the potential conversion technologies for resources or waste, in a context submitted to seasonal variations. Following the developed methodology, the system design is a problem that can be solved by extending the process design techniques to larger systems including not only the process flowsheet design of detailed conversion technologies, but also the choice of the raw materials and resources, the supply chains and the waste management and recycling options.

5.1.1   Urban systems superstructure

The general concept described earlier for the extended action system and supply chain synthesis in section 2.3.1 and Figure 2.10 is taken as a basis to generate the superstructure that is used for the design and synthesis of an urban system. In this particular context, it can be subdivided in five different subsystems:

1) the available resources. These can be indigenous – e.g. biomass or geothermal energy – in which case they are limited, or imported – e.g. natural gas or oil – in which case they can be limited or not, depending on the problem to be solved.

2) the conversion technologies. These are necessary to convert the resources into final energy services or intermediate products – e.g. gasification of biomass or cogeneration engines – or to treat the waste – e.g. wastewater treatment plant.

3) the services to be supplied. Though only energy services are considered in the present case study, it is as well possible to include other types of services – e.g. drinkable water or food to be supplied in a given area.

4) the waste to be treated – e.g. municipal solid waste or wastewater.

5) the transfer networks. These are used when the services can not directly be delivered to the user without an intermediate – e.g. district heating network.

Following the methodology presented in section 2.3, each unit contained in these subsystems includes associated mass and energy flows, operating costs (CO_u), investment costs (CI_u), and environmental impacts (I_u) extracted from the disaggregated data of the LCI database. All these parameters are calculated for a nominal size $(f_u = 1)$. Each unit is assigned as well a minimal and maximal sizing factor $(f_{min}$ and $f_{max})$, which depends on if the unit is a process $(f_u = 1)$, a utility with unlimited use $(f_{max} =$ unlimited$)$ or with limited use $(f_{max} =$ limited$)$. The units contained in the subsystems can be adapted depending on the case study, and are either detailed models generated using flowsheeting software, or average technologies extracted from the LCI database. The example of such a superstructure is displayed in Figure 5.1, with both detailed models and average technologies from the LCI database ecoinvent®.

Following the developed methodology for the extraction of average technologies from the LCI database, as presented in section 2.3.2, the conversion technologies are s-type processes, with an incremental impact, and the resources are r-type processes, with a cumulated impact. Thus, there is no risk of double-counting the emissions while synthesizing the supply chains for the overall system.

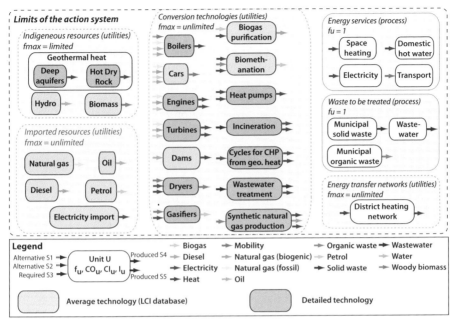

Figure 5.1 Example of a superstructure for urban energy systems synthesis (adapted from Gerber et al. (2013)).

5.1.2 Case study description

The city of La Chaux-de-Fonds, in Switzerland, is taken as an application case study. It is located at 1000m of altitude in the Jura mountains and has a population of approximately 40 000 inhabitants. A district heating network is already existing, with an installed capacity of 47 MW_{th}. The heat requirements are presently supplied in priority by a Municipal Solid Waste Incineration (MSWI) plant, which has a steam network that allows for the cogeneration of electricity and district heating. The remaining heat requirements are supplied by a wood boiler and a natural gas boiler.

Goal and scope definition

Accounting for the existing facilities, the objective is to redesign and optimize the urban energy system of La Chaux-de-Fonds in a future perspective, considering the different energy services to be supplied and the waste to be treated, as well as the available indigenous resources. Both economic and environmental criteria have to be considered, in order to compare the potential new configurations with the present situation. The optimal value that has to be fixed for a CO_2 tax in order to favor economically the system configurations reducing the environmental impact is as well considered. The superstructure of Figure 5.1 is used, and displays a list of the services and waste to be considered, as well as of the available indigenous resources, the potential imported resources and the potential energy conversion technologies that can be used. The environomic models of these different components are briefly presented in section 5.2 below.

The functional unit of the considered system, both for evaluating the economic and environmental performance, is the yearly energy services to be supplied to the inhabitants of the city, as well as the yearly waste treatment and disposal. In all economic calculations of the present chapter, the currency exchange rate is 1.2 CHF/€ (01.05.2012).

5.2 Models description

5.2.1 Energy services

The energy services to be supplied include space heating and hot water, electricity and mobility.

Space heating and hot water

The seasonal requirements in space heating and hot water have been characterized for a residential area of La Chaux-de-Fonds by Mégel (2011), using the methodology of Girardin et al. (2010). These data are used as a basis to calculate the total heating requirements for the buildings connected to the district heating network with an installed capacity of 47 MW$_{th}$. Only the requirements for the district heating network are considered in the present case study, and not the individual domestic heating. Figure 5.2 displays these seasonal requirements in space heating and hot water.

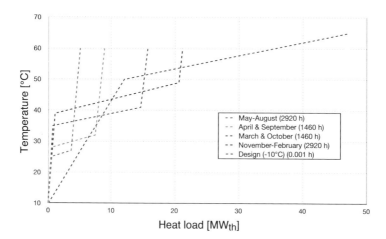

Figure 5.2 Seasonal demand profiles in space heating and hot water for the buildings connected to the district heating network of La Chaux-de-Fonds (adapted from Gerber et al. (2013)).

The seasonal requirements are used as a basis to define the periods considered in the present case study, their duration being displayed in Figure 5.2. During summer, the MSWI plant is shut down one month for revision (Mégel (2011)). Thus, the month of June is considered as a separate period from May, July and August, with the same heating requirement.

Electricity

The monthly electricity consumptions for overall Switzerland (OFEN (2010c)) are used to estimate the electricity requirements per capita corresponding to each period for the present case study. The seasonal demand in electricity per capita and accounting for the losses of the grid is displayed in Table 5.1.

Table 5.1 Seasonal electricity requirements per capita for the case study.

Period	Demand, in kW_e/cap
May, July & August	0.886
June	0.903
April & September	0.931
March & October	1.020
November-February	1.110
Design (peak load)	1.4

Mobility

According to the Swiss Federal Office of Statistics (OStat (2010)), the mobility requirements in Switzerland are currently around 11 392 pkm[1] per year and per capita. This value is used to calculate the overall mobility requirement for the case study. No seasonal variation is assumed for the mobility service.

5.2.2   Waste to be treated

The waste to be treated considered in the present case study include municipal solid waste (MSW), wastewater and organic waste (OW). Energy services can be potentially generated from all theses types of waste.

Municipal Solid Waste

From the data available for the MSWI plant, 55 000 tons are incinerated each year, representing the MSW generated by the city of La Chaux-de-Fonds and the surrounding area (Cridor (2011)). They are treated in the MSWI plant, which has a steam network to produce Combined Heat and Power (CHP). Since

---

[1]pkm: person·kilometer

the MSWI plant is shut down one month in June, the total amount is divided between the remaining periods (11 months), considering no seasonal variation.

Organic waste

The MSWI plant includes as well facilities for the treatment of 3500 tons of organic waste each year, which are currently composted (Cridor (2011)). This amount could be used to satisfy a part of the requirements in energy services, and is therefore included in the case study. No seasonal variation is assumed.

Wastewater

From the data available in Muller et al. (2008) and assuming no seasonal variation, the design value taken for the wastewater entering a wastewater treatment plant (WWTP) is equal to 300 m^3 per year and per capita.

5.2.3   Indigenous resources

According to the developed methodology applied to an urban system, the different indigenous resources available on the considered territory have to be identified and their potential quantified. In the present case study, four different indigenous renewable resources are considered: residual wood from forest industry, geothermal resources, considering both deep aquifers and Hot Dry Rock, and hydraulic energy. Each one of these resources has an associated operating cost and a cumulated life cycle inventory accounting for all upstream emissions, calculated by a detailed model or extracted from ecoinvent® when average market data are used. The ecoinvent® equivalences and the references for the LCI models are given in Annex A.3.

Woody biomass

According to Steubing et al. (2010), the sustainable biomass potential in Switzerland for forest energy wood, industrial wood residues and wood from landscape maintenance is equal to 34.9 PJ$_{th}$ per year. Brought back to the population of La Chaux-de-Fonds, this gives a total yearly potential of 49 086 MW$_{th}$ per year. The assumed price – i.e. operating cost – for biomass is 0.05 €/kWh, according to Mégel (2011). The investment cost associated with the exploitation of the woody biomass is assumed to be already accounted for in the operating cost.

Regarding the environmental impacts, the average market technology is assumed to be representative, and no detailed model is used for the harvesting of the woody biomass. The wood chips are assumed to be a mix of hard wood and soft wood chips, calculated in m^3, with an energy density of 3294 MJ/m^3 (Werner et al. (2003)).

No seasonal variation is assumed in the wood availability, but it can be distributed among the different considered periods.

Geothermal energy

For the geothermal potential of the case study, two types of resources are considered: deep aquifers and Hot Dry Rock (HDR), the latter being exploitable as an Enhanced Geothermal System (EGS).

The geothermal potential for deep aquifers has been quantified in a technical report for the development of geothermal energy in the region of Neuchâtel (Working group PDGN (2010)). It appears from this report that for the city of La Chaux-de-Fonds, two resources could potentially be exploited for geothermal applications, in the geological layers of the Dogger, at 950m, and of the Muschelkalk, at 1450m. The potential integration of these two resources for supplying heat to the district heating network with current operating conditions via heat pumps has been studied in another technical report (Gerber and Maréchal (2011)).

Since the economic and thermodynamic performances of the Muschelkalk are expected to be better than the ones of the Dogger, only the former is considered in the present case study. Its expected temperature is 51°C, its reinjection temperature 30°C, its expected mass flow rate is 15 kg/s, and the available thermal power 1.1 MW$_{th}$. For evaluating the investment and the operating costs, data from the report of the Working group PDGN (2010) and from Gerber and Maréchal (2011) are used.

For the evaluation of environmental impacts, the LCI model from Gerber and Maréchal (2012c) described in section 4.2.2 of Chapter 4 is adapted to the deep aquifer. This means that for the number of wells $n_w = 2$, and that the impacts from the site exploration, drilling, casing, cementation and end-of-life are kept and adapted to shallower wells, while the ones from the reservoir enhancement and stimulation are removed.

It is possible to consider the future integration of an EGS in the urban energy system of La Chaux-de-Fonds, either for electricity production or for CHP. The data from the Working group PDGN (2010) considering only aquifers and not deeper HDR resources, the data from Gerber and Maréchal (2012a,c) described earlier in Chapter 4 are used. The five typical optimal configurations of subsection 4.3.3 of Chapter 4 are considered in the superstructure with their associated investment costs, operating costs and LCI. This corresponds to four potential construction depths for the EGS: 6000m (Kalina cycle for CHP), 6500m (ORC for electricity production), 8000m (flash system for CHP) and 9500m (flash with bottoming ORC for electricity production and ORC with draw-off for CHP).

Hydraulic energy

Hydro-electricity production represents around 56% of the Swiss electricity production (OFEN (2010c)) and has thus to be considered in the base mix of Swiss electricity. Using the data from OFEN (2010c), the hydro-electricity production can be estimated to 0.536 kW_e in average per capita. This value is taken for the case study to define a threshold for the maximal electricity production that can be allocated to the population of La Chaux-de-Fonds. The operating costs from hydro-electricity import are assumed to be 0.16 €/kWh$_e$, which is an average price for the Swiss electricity market (OFEN (2010c)).

For the evaluation of environmental impacts, the average LCI for the production of hydro-electricity in Switzerland is assumed to be representative (Dones et al. (2007)).

5.2.4   Imported resources

Since the waste to be treated and the indigenous resources may not be sufficient to satisfy the yearly overall requirements in energy services, imported resources have to be considered as well. For the present case study, natural gas, light fuel oil, petrol, diesel and electricity mix from the UCTE grid are considered as potential imported resources. Table 5.2 summarizes the operating costs associated with these resources. No variation over time is assumed for these prices. Since these resources are imported, the investment costs are not accounted for in the economic model, and are therefore indirectly included in the operating costs. Nuclear power, and thus uranium, is not considered in the imported resources, due to the future perspective of the present case study in a Swiss context, where nuclear power is planned to be abandoned by 2034. No seasonal variation is assumed for all imported resources.

Table 5.2 Prices of the imported resources for the case study.

Resource	Price
Natural gas	0.078 €/kWh
Light fuel oil	0.083 €/kWh
Petrol	1.88 €/kg
Diesel	1.75 €/kg
Electricity UCTE mix	0.16 €/kWh$_e$

Since the average market technology is assumed to be representative for these imported resources, the equivalences for the LCI models are all taken from ecoinvent[®] in Dones et al. (2007), and are given in Annex A.3.

5.2.5   Conversion technologies

Different conversion technologies are included in the superstructure of the urban energy system of La Chaux-de-Fonds, in order to treat the waste and to convert the resources in intermediate products and useful energy services. A short description of the environomic model of each conversion technology considered for the present case study is presented in the paragraphs below.

A graphical representation of the simplified model for each energy conversion technology embedded in the superstructure is available in Annex D. The necessary data for solving the system design in the MILP subproblem are detailed. This includes the associated major mass and energy streams, as well as the specific operating costs $(\dot{CO}_u)$ and impacts $(\dot{I}_u)$ per nominal size of the unit at nominal operating conditions. Both the operating costs and the impacts displayed concern only s-type units, and not all the upstream inherited emissions and costs from the resources, in agreement with the methodology to avoid double-counting. The ecoinvent® equivalences or the references for the LCI models are given in Annex A.3.

Biomethanation

Biomethanation is used for the conversion of organic waste to biogas, which can be burned then for CHP either inside an engine or a turbine, or sent to a purification unit to be converted to Synthetic Natural Gas (SNG). The process of biomethanation requires both electricity and heat. The physical model is based on the average technology of ecoinvent® (Jungbluth et al. (2007)). The economic data for the investment and operating costs are taken from Gassner et al. (2011). For the evaluation of environmental impacts, the LCI data are taken from Jungbluth et al. (2007).

Boilers

Boilers are used to burn woody biomass, light fuel oil, natural gas or SNG to produce heat for the district heating network or conversion technologies requiring heat. The physical models are taken from Fazlollahi and Maréchal (2013). They include economic data for the investment and operating costs and for biogenic or fossil CO_2 emissions. For the remaining emissions of the LCI, the data from Dones et al. (2007) of ecoinvent® are used.

Cars

Cars are used to supply individual mobility, using diesel, petrol, natural gas or SNG as a fuel. The physical models and LCI are directly extracted from the average market technology available in ecoinvent® (Dones and Bauer (2007)). The investment costs and maintenance costs are not accounted for, since they are assumed to be independent of the fuel type.

Cycles for the conversion of heat from EGS

The five typical optimal configurations detailed in the results of Chapter 4 are included in the superstructure, with their associated investment costs, operating costs and LCI. This corresponds to five different conversion technologies either for single electricity production or for CHP: a Kalina cycle for CHP (with an EGS at 6000m), a supercritical ORC using R134a for single electricity production (with an EGS at 6500m), a double-flash system for CHP (with an EGS at 8000m), a single-flash system with a bottoming ORC using iso-butane for single electricity production (with an EGS at 9500m), an ORC using cyclo-butane with an intermediate draw-off for CHP (with an EGS at 9500m). The configuration selection is treated as a decision variable, since it is considered that only one EGS can potentially be built for the city of La Chaux-de-Fonds, given the size of the city.

Dryers

The dryers are used to dry woody biomass prior gasification. Two physical models are considered, based on the work of Fazlollahi and Maréchal (2013): air drying and steam drying. These models include as well economic data for the investment and operating costs. The LCI for the dryers is taken from Gerber et al. (2011a) (see Chapter 3).

Engines

The engines are used for CHP from natural gas, SNG or biogas. The physical models are taken from Fazlollahi and Maréchal (2013). They include as well economic data for the investment and operating costs. For the LCI, the data from Dones et al. (2007) and Jungbluth et al. (2007) of ecoinvent® are used.

Gasifiers

The gasifiers are used to convert woody biomass to producer gas that can be burned then for CHP either in an engine or a turbine. Four physical models are considered, based on the work of Fazlollahi and Maréchal (2013): air gasification and steam gasification, both for the production of gas for the engine or for the turbine. These models include as well economic data for the investment and operating costs. The LCI for the gasifiers is taken from Gerber et al. (2011a) (see Chapter 3).

Heat pumps

Two models of heat pumps are included in the superstructure, with different temperature levels: high temperature heat pump (HT HP) and low temperature heat pump (LT HP). The working fluid is R134a. The physical model and economic data including investment and operating costs are taken from Gerber and Maréchal (2011) and Gerber and Maréchal (2012a). For the LCI, the data from Dones et al. (2007) are used.

Municipal Solid Waste Incineration

A simplified model of the MSWI plant is presented in Gerber and Maréchal (2011), based on the operation data of the plant. 17 MW_{th} of steam are produced from the combustion of waste, from which 16.5 MW_{th} are available for the combined production of district heating and of electricity. The electricity is produced using a condensation turbine with a maximal power of 5 MW_e and a minimal power of 1 MW_e, with an electrical efficiency of 31% (Mégel (2011)). This allows for performing the energy balance and for calculating the remaining heat available for the district heating as a function of the electricity production.

Since the facility is already existing, the investment cost is not accounted for. The operating costs are equal to 0.03 €/kWh_{th} supplied to the district heating, with the current operating conditions (Mégel (2011)). For the LCI, the data from Doka (2007) are used.

Purification of biogas

The purification and pressurization allows for converting the biogas from the biomethanation to SNG of a sufficient quality to be injected in the gas grid. This process requires electricity. The model for purification and pressurization is based on the average technology of ecoinvent® (Jungbluth et al. (2007)). The economic data for the investment and operating costs are taken from Gassner et al. (2011). The LCI data are taken from Jungbluth et al. (2007).

SNG production

The thermochemical conversion of woody biomass followed by its methanation and purification allows for the combined production of SNG, electricity and heat. The physical and economic model, including both the investment and the operating costs, is based on the work of Gassner (2010), which has already been used in Chapter 3. For the present application case study, only the technology using indirect steam-blown gasification (FICFB) with membranes for purification is considered. The LCI data are taken from Gerber et al. (2011a) (see Chapter 3).

Turbine

The turbine is used for the combined heat and power production from biogas. The physical model is based on the work of Fazlollahi and Maréchal (2013). It includes as well economic data for the investment and operating costs. For the LCI, the data from Primas (2007) of ecoinvent® are used.

Wastewater Treatment Plant

The wastewater treatment plant includes a facility for sludge digestion and its conversion to biogas, which allows for on-site combined heat and power production. The electricity is first used for the internal consumption of the WWTP. If additional electricity is produced, it is exported to the grid. The physical model is based on the work of Descoins et al. (2012), which includes as well economic data for the operating costs. Since the wastewater treatment plant is an existing facility, the investment cost is not accounted for. For the LCI, the data from Doka (2007) of ecoinvent® are used.

5.2.6 District heating network

The existing district heating network allows for supplying heat to the users, the demand being displayed in Figure 5.2. When compared with the current operating conditions of the district heating network, Mégel (2011) demonstrates that lowering the return temperature would allow for decreasing significantly the operating costs of the overall system. Thus, the minimal possible return temperature – 20°C lower than the current operating conditions – has been assumed for the present case study. The return and supply temperatures depending on the season are displayed in Table 5.3.

Table 5.3 Return and supply temperatures for the district heating network of La Chaux-de-Fonds.

Period	Supply [°C]	Return [°C]
May, July & August	90	38
June	90	38
April & September	92	39
March & October	96	41
November-February	99	43
Design (peak load)	112	45

Since the district heating network is already existing, the costs and the impacts of its construction are not accounted for. It is used as a constraint for the heat supply from the different technologies, avoiding the direct exchanges with the heat demand of the users. The methodology from Becker and Maréchal (2012a) for the heat exchanges with restriction is applied to account for this at the energy integration step.

5.2.7 Current situation

Since the present case study is dealing with an existing system in which the integration of emerging technologies and the optimization of operating conditions are studied, the actual economic and environmental performances of the system have to be assessed. The performances of the new system configurations obtained by environomic optimization can then be compared with a reference scenario. Therefore, the actual operating costs and environmental impact have to be calculated.

According to the monthly data available in Mégel (2011) for the MSWI plant operation, 57% of the district heating requirements are supplied by the MSWI plant, while 13% are supplied by a wood boiler and 30% by a natural gas boiler. Regarding the electricity consumption, 7% is supplied by the MSWI plant. The remaining demand is considered to be supplied by the average Swiss electricity mix of the consumer, including the imports (Dones et al. (2007)). The mobility is assumed to be supplied by fossil fueled cars with a share of 85% for petrol cars and 15% for diesel cars (Dones and Bauer (2007)).

Figure 5.3 shows the calculated yearly CO_2-equivalent emissions per capita (4.0 tons) and the yearly total operating costs (95.3 millions of €) for the current situation. The flowchart diagrams representing the current seasonal operation of the system are available in Annex E. As it can be seen, the environmental impacts and the operating costs of the current situation are mainly caused by the fossil fuel consumption for the transport and by the average Swiss electricity mix. The impact of the natural gas boiler for supplying the district heating requirements is relatively low, since a high share of it is already supplied by the MSWI plant.

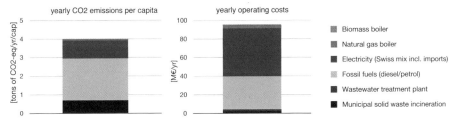

Figure 5.3 Environomic performance of the current situation in La Chaux-de-Fonds: yearly CO_2 emissions per capita (left) and yearly total operating costs (right).

5.3 Environomic optimal configurations

The goal of the optimization is to identify the optimal configurations for the urban energy system of La Chaux-de-Fonds and its waste treatment system, considering the seasonal variations, the potential indigenous and imported resources, the waste recycling and the potential conversion technologies. The supply chain synthesis is thus included in the optimization problem, and an extended decision perimeter of the process flowsheet is considered, including the resources, waste and average technologies. The objective for the MILP slave sub-problem is the minimization of the operating costs (Equation (2.29)). This includes as well the costs of the life-cycle CO_2 equivalent emissions, expressed as the Global Warming Potential on a 100-year time-horizon and multiplied by a CO_2 tax (see Equation (2.29)), which is given as one of the decision variables of the MINLP master multi-objective optimization problem.

5.3.1   Multi-objective optimization problem

Three independent optimization objectives are selected:

1) The investment costs, to be minimized:

$$Min \; CI_{tot} = \sum_{t=1}^{n_t} max(CI_t(\mathrm{f_t}, x_d, c_I))_{n_p} + \sum_{r=1}^{n_r} max(CI_r(\mathrm{f_r}, x_d, c_I))_{n_p} \quad (5.1)$$

where n_t and n_r are the number of technologies and of resources, respectively, embedded in the superstructure, CI_t and CI_r is the investment cost linked with technology t or resource r, for which the maximum value is retained over all the periods n_p, $\mathrm{f_t}$ and $\mathrm{f_r}$ are the utilization factors of the technologies and of the resources, respectively, and are the decision variables of the MILP slave sub-problem, x_d are the decision variables of the MINLP master problem, and c_I is the cost of the environmental impact, here equivalent to the CO_2 tax.

2) The annual operating costs, to be minimized:

$$Min \; CO_{an} = \sum_{p=1}^{n_p} t_p \cdot (\sum_{t=1}^{n_t} \dot{CO}_t(\mathrm{f_t}, x_d, c_I) + \sum_{r=1}^{n_r} \dot{CO}_r(\mathrm{f_r}, x_d, c_I)) \quad (5.2)$$

where n_p is the number of independent periods considered, t_p is the operating time associated with period p, $\dot{CO}_t$ is the operating cost of technology t and $\dot{CO}_r$ is the total operating cost of resource r.

3) The annual life-cycle CO_2-equivalent emissions per capita, to be minimized:

$$Min \; I_{an,cap} =$$

$$\frac{\sum_{t=1}^{n_t} (max(IC_t)_p + max(IE_t)_{n_p} + \sum_{p=1}^{n_p} t_p \cdot \dot{IO}_t)(\mathrm{f_t}, x_d, c_I)}{n_{cap} \cdot t_{yr}}$$

$$+ \frac{\sum_{r=1}^{n_r} (max(IC_r)_p + max(IE_r)_{n_p} + \sum_{p=1}^{n_p} t_p \cdot \dot{IO}_r)(\mathrm{f_r}, x_d, c_I)}{n_{cap} \cdot t_{yr}} \qquad (5.3)$$

where IC_t and IC_r is the construction impact of technology t and resource r, respectively, IE_t and IE_r is its end-of-life impact, for which the maximum value is retained over all the periods n_p, $\dot{IO}_t$ and $\dot{IO}_r$ is its operating impact, n_{cap} is the population of the considered urban area, here 40 000, and t_{yr} is the system lifetime, here assumed to be 25 years.

In order to ensure that the whole space of potential configurations is explored by the evolutionary algorithm (Molyneaux et al. (2010)), the selection of each one of the different technologies embedded in the superstructure described in Figure 5.1 for solving the MILP slave sub-problem is given as a decision variable of the MINLP master problem. All the decision variables are given in Table 5.4. Some of these variables are generated using a multi-period strategy, in order to adapt the system operating conditions to the seasonal variations.

5.3.2   Final performance indicators

The optimal system configurations have to be compared in terms of economic and of environmental performance with the current situation, or reference scenario, in order to identify the most suitable ones.

For the economic performance, the simple payback period is used and is expressed as:

$$t_{pb} = \frac{CI_{an} \cdot t_{yr}}{CO_{an,ref} - CO_{an}(c_I)} \qquad (5.4)$$

where CI_{an} are the annualized investment costs of the configuration to be compared (see Equation (4.16)) with an interest rate ir of 6% and a system lifetime t_{yr} of 25 years, $CO_{an,ref}$ are the yearly operating costs of the reference scenario, estimated to 95.3 millions of €/yr (see Figure 5.3), and CO_{an} are the yearly operating costs of the configuration. These can be recalculated as a function of a varying CO_2 tax to study its effect on the most suitable economic configurations and to determine its optimal value by comparison with the environmental most suitable configurations.

For the environmental performance, the relative reduction in the impact, here the CO_2-equivalent emissions, is used and is expressed as:

$$\delta_I = \frac{I_{an,cap,ref} - I_{an,cap}}{I_{an,cap,ref}} \qquad (5.5)$$

where $I_{an,cap,ref}$ is the environmental impact of the reference scenario, estimated to 4.0 tons per year and per capita of CO_2-equivalent emissions (see Figure 5.3), and $I_{an,cap}$ is the environmental impact of the configuration to be compared.

Table 5.4 Decision variables used for the multi-objective optimization of the case study of La Chaux-de-Fonds for urban systems.

Name	Type of variable	Multi-period	Range	Unit
CO_2 tax c_I	continuous	no	[0; 200]	€/tCO_2-eq
Electricity production from MSWI plant	continuous	yes	[1000; 5000]	kW$_e$
Biomass usage over period p	integer	yes	[0; 1]	–
Evaporation temperature of HT HP	continuous	yes	[5; 40]	°C
Evaporation temperature of LT HP	continuous	yes	[5; 40]	°C
Condensation temperature of HT HP	continuous	yes	[30; 80]	°C
Condensation temperature of LT HP	continuous	yes	[30; 80]	°C
EGS configuration choice[a]	integer	no	[0; 5]	–
Selection of biomethanation of OW	integer	no	[0; 1]	–
Selection of diesel car	integer	no	[0; 1]	–
Selection of petrol car	integer	no	[0; 1]	–
Selection of fossil NG car	integer	no	[0; 1]	–
Selection of biogenic SNG car	integer	no	[0; 1]	–
Selection of air biomass dryer	integer	no	[0; 1]	–
Selection of steam biomass dryer	integer	no	[0; 1]	–
Selection of biogas engine	integer	no	[0; 1]	–
Selection of fossil NG engine	integer	no	[0; 1]	–
Selection of biogenic SNG engine	integer	no	[0; 1]	–
Selection of air gasification of biomass for engine	integer	no	[0; 1]	–
Selection of steam gasification of biomass for engine	integer	no	[0; 1]	–
Selection of air gasification of biomass for turbine	integer	no	[0; 1]	–
Selection of steam gasification of biomass for turbine	integer	no	[0; 1]	–
Selection of biogas turbine	integer	no	[0; 1]	–
Selection of OW biogas purification to SNG	integer	no	[0; 1]	–
Selection of thermochemical wood conversion to SNG	integer	no	[0; 1]	–

[a] 0: no EGS, 1: 6000m EGS, Kalina cycle, 2: 6500m EGS, ORC-s with R134a, 3: 8000m EGS, 2F, 4: 9500m EGS, 1F & ORC with iso-butane, 5: 9500m EGS, ORC-d with cyclo-butane.

5.3.3 Results and discussion

Pareto curve

Figure 5.4 shows the Pareto curve obtained for the tri-objective environomic optimization for the urban system of La Chaux-de-Fonds.

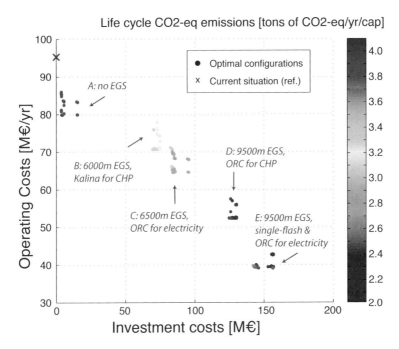

Figure 5.4 Pareto curve obtained by the tri-objective environomic optimization. The current situation (reference scenario) is as well displayed on the graph (adapted from Gerber et al. (2013)).

Five clusters are identified on the Pareto curve, each one of them corresponding to the selection or not of one of the EGS configurations embedded in the super-structure. The curve shows clear trade-offs between the investment costs and the yearly operating costs, and between the investment costs and the yearly life cycle CO_2 emissions per capita. Inside all the clusters, there is as well a trade-off between the CO_2 emissions and the two other economic objectives, though the variations in terms of economic and environmental objectives are less important than between the clusters. This is illustrated by Figure 5.5, which shows the details of cluster E with a more precise scale for the CO_2 emissions.

As illustrated by Figure 5.5 for cluster E, the introduction of an environmental objective in the optimization procedure leads to consider configurations that would be suboptimal in a pure economic optimization and that would thus not be considered in the final optimal configurations of the system. Indeed, though suboptimal in terms of operating or investment costs, these environomic system

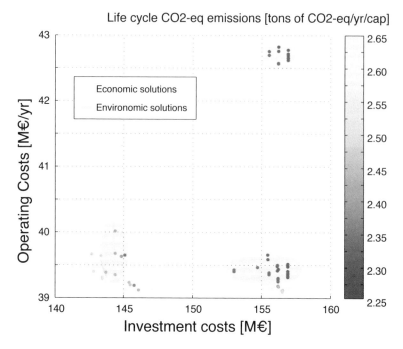

Figure 5.5 Detail of cluster E of the Pareto curve on Figure 5.4 (adapted from Gerber et al. (2013)).

configurations lead to a lower environmental impact. Therefore, in the following subsection, one economic and one environomic configuration are discussed in details for each cluster of the Pareto curve.

Typical configurations

Figure 5.6 displays the investment costs (Figure 5.6(a)), the operating costs (Figure 5.6(b)) and the life-cycle CO_2 emissions (Figure 5.6(c)) associated with 10 typical configurations of the Pareto curve of Figure 5.4. For each cluster of the Pareto curve (A,B,C,D,E), one pure economic configuration (1) and one environomic configuration (2) are analyzed. The operating costs and CO_2 emissions of the current situation (Ref) are as well displayed for comparison. The detailed results of these configurations are available in Annex E. A flowchart diagram representing the seasonal operation of the system per capita is made for each period and for each configuration.

The investment costs are dominated by the geothermal technologies – i.e. the deep aquifer and the EGS – that have important costs associated with the drilling of wells. The investment of a wood-to-SNG conversion unit is as well significant for the configurations for which this technology is selected. When compared with the current situation, all the optimal configurations lead to a reduction in the system yearly operating costs. These are dominated by the electricity costs either from hydro-electricity or from imports of UCTE

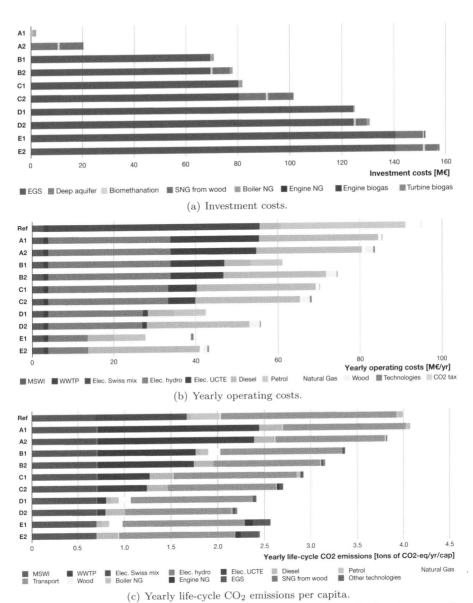

(a) Investment costs.

(b) Yearly operating costs.

(c) Yearly life-cycle CO_2 emissions per capita.

Figure 5.6 Comparison of typical optimal configurations from the Pareto curve of Figure 5.4 . (a) Investment costs (b) Yearly operating costs (c) Yearly life-cycle CO_2 emissions per capita (adapted from Gerber et al. (2013)).

electricity mix, and by the costs of fossil fuels imports for transport. The contribution of the CO_2 tax is low in all the cases, since the highest CO_2 tax selected in the optimal configuration is equal to $c_I = 2$ €/ton of CO_2-eq. This is due to the inclusion of the technology selection in the decision variables of the multi-objective optimization master problem, which ensures that the configurations having a good environmental performance are selected even in the case when the environmental tax is low. All the optimal configurations, except A1, lead to a reduction in the environmental impact. The CO_2 emissions are dominated by the electricity UCTE mix import and by the fossil CO_2 emissions due to transport.

The major difference between the economic and the environomic configurations is the use of a wood-to-SNG conversion unit, of a biomethanation unit for organic waste, and, in some of the cases, the exploitation of a deep aquifer for geothermal energy. When comparing the environomic solution with the economic solution of the same cluster, this has the consequence to increase the total investment costs, but to decrease the CO_2 emissions by substituting a part of the fossil resources used for transport, heat and electricity, and of the UCTE electricity mix import. Both economic and environomic configurations use EGS at different depths, with different cycles, either with or without heat production in addition to the electricity production.

The detailed analysis of each cluster below provides more insight on the technologies and resources reducing the costs and the impacts, and on the strategies that could be adopted for operating the urban energy system of this application case study.

1) Cluster A: no EGS. This is the only cluster not using an EGS, having thus the lowest investment costs. Configuration A1 uses only a biomethanation unit for organic waste and a natural gas boiler in addition to the existing facilities. This configuration is therefore almost equivalent to the current situation. The only significant differences are:

   - the switch from petrol to diesel in the fossil fuels consumption, which reduces the operating costs.
   - the switch from the actual Swiss mix, which includes hydro-electricity and nuclear power, to a future mix including hydro-electricity from Switzerland and UCTE mix import.
   - a reduction in the natural gas consumption, due to the optimization of the ratio between heat and power produced by the MSWI plant.

   The overall impact is slightly increased when compared with the current situation. Though the impact from transport emissions is reduced by the use of diesel instead of petrol, this is more than compensated by the import of UCTE mix, which has higher specific CO_2 emissions (0.516 kgCO_2-eq/kWh$_e$) than the current Swiss mix of the consumer including the imports (0.121 kgCO_2-eq/kWh$_e$). The investment in a wood-to-SNG conversion unit, producing as well heat and electricity, and in a deep aquifer for the use of geothermal heat (see configuration A2) reduces the emissions from transport, and as well slightly from heat and electricity. These reductions are sufficient to make this configuration as well slightly beneficial in terms of environmental impact when compared with the current situation (4% reduction).

2) Cluster B: EGS at 6000m, using a Kalina cycle for CHP. The use of this EGS configuration allows for reducing significantly the import of UCTE electricity mix, which decreases the operating costs. The environmental impacts are as well reduced, between 16 and 21%, depending on if an investment is made or not in a wood-to-SNG conversion unit. No investment is made in a deep aquifer in the environomic configuration B2, since the EGS is used for CHP. Therefore, enough heat is available to supply the district heating requirements year-round in addition to the MSWI plant. This is as well the case for the other EGS configurations with CHP (cluster D), but not for the configurations with an EGS for single electricity production (cluster C and E), where for some of them a deep aquifer is selected.

3) Cluster C: EGS at 6500m, using a supercritical ORC with R134a for single electricity production. The increased electricity production from a deeper EGS than for cluster B and the use of an ORC for single electricity production reduces further the operating costs and the environmental impacts from the import of UCTE electricity mix. Depending on if a wood-to-SNG conversion unit and a deep aquifer are used or not, the impact reduction is between 27% and 32% compared with the current situation. Since the EGS is used for electricity production only and that the heat from the MSWI is not sufficient, the environomic configuration C2 uses a deep aquifer in winter to supply part of the district heating requirements.

4) Cluster D: EGS at 9500m, using an ORC with cyclo-butane and an intermediate draw-off for CHP. The EGS is deeper than for cluster B and C, and leads therefore to an increased electricity production. This allows for reducing consequently the import of UCTE mix and of hydro-electricity, the production of electricity from the MSWI and the EGS being sufficient during the summer periods. Therefore, the operating costs and the impacts are both reduced. The environomic configuration D2, with a wood-to-SNG conversion unit and a biomethanation unit in addition to the EGS for CHP, has the best performance of all in terms of environmental impact, with yearly CO_2-equivalent emissions of 2.2 tons per capita, which corresponds to a reduction of 45% compared to the current situation.

5) Cluster E: EGS at 9500m, using a single-flash and a bottoming ORC with iso-butane for single electricity production. The EGS, at the same depth than cluster D, is exclusively used for electricity production, which makes the city almost completely autonomous for this energy service – except for a small production from natural gas engines – considering as well hydro-electricity as a local resource. This reduces further the operating costs, the economic configuration E1 being the best one with respect to this indicator, with a reduction of more than 50% when compared with the current situation. However, this cluster has a slightly less good environmental performance than cluster D, mostly due to the fossil CO_2 emissions from the EGS, since one of the conversion technologies is a flash system, which has

direct emissions into the atmosphere. Though much less important than the emissions from fossil sources, this indicates that the fossil CO_2 emissions from geothermal energy may have an influence on the final choice of the system configuration if environmental criteria are considered.

Though the operating costs and environmental impacts linked with the district heating and with the electricity can be potentially both reduced in an important way by relying on waste treatment technologies and on local resources, the costs and the impacts linked with the transport service remain high even when the full potential of woody biomass is converted into SNG and used in cars. This highlights therefore the importance of accounting for the availability of local renewable resources and waste.

Four important aspects linked to the application of the methodology to the case study can be identified:

- the seasonal character of the energy services and thereby of the system operation.
- the optimal supply chain for resources and waste, and their allocation for supplying energy services – i.e. the optimal pathways.
- the selection of the technologies embedded in the superstructure.
- the competition or the synergies between the different conversion chains.

The developed methodology allows for accounting for the seasonal variation in the energy services demand – i.e. electricity and heating – for the variations in the operation of the facilities – i.e. shut down of the MSWI plant in June – and to adapt the technology selection and the operating conditions to these constraints. This aspect is illustrated by the configuration E1 in Figure 5.7, for which the system operation per capita is represented for three different periods. The technologies or resources presenting a seasonal variation are highlighted with their material and energy flows in color, while the ones with constant operation are shaded and their material and energy flows are displayed in grey.

As displayed in Figure 5.7, the share between electricity and district heating from the MSWI plant, the heat usage from the WWTP, the consumption of hydro-electricity, the usage of the deep aquifer and of the natural gas engine are adapted to the energy service requirements and if the MSWI plant is operating or not.

During the first summer period – May, July and August – the district heating requirement is rather low and can be fully satisfied by the waste treatment facilities (see Figure 5.7(a)). For the MSWI plant, the emphasis is thus put on the electricity production, and only a small import of hydro-electricity is required, representing 4% of the electricity consumption.

During the second summer period – June – the MSWI plant is shut down for revision. The district heating requirement being still low, more heat available

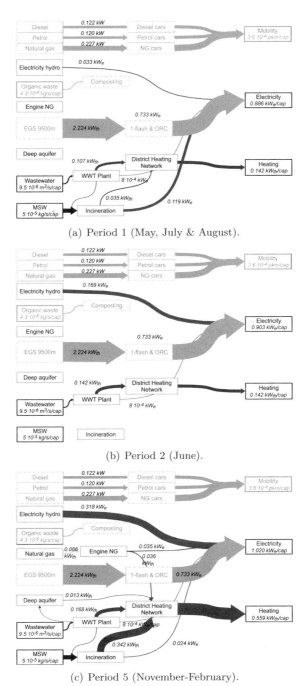

(a) Period 1 (May, July & August).

(b) Period 2 (June).

(c) Period 5 (November-February).

Figure 5.7 Examples of seasonal operation per capita for configuration E1: (a) Period 1 (May, July & August) (b) Period 2 (June) (c) Period 5 (November-February) (adapted from Gerber et al. (2013)).

from the WWTP plant is used to compensate the non-operation of the MSWI plant (see Figure 5.7(b)). The district heating requirement is thus still supplied only with local resources from waste treatment. For the electricity production, more hydro-electricity is consumed to compensate the shut down of the MSWI plant, the potential of hydro-electricity per capita (0.536 kW$_e$/cap) being not fully used.

During the winter period – November to February – the district heating requirement increases strongly – 400% compared to the summer requirement – while the electricity requirement increases slightly – 25% compared to the summer requirement. Therefore, the operating conditions of the MSWI plant are adapted to favor the heat production over the electricity production (see Figure 5.7(c)). The WWTP supplies as well slightly more heat, but is limited by the temperature of the district heating network, which is higher in winter than in summer. Part of the low-temperature heat is therefore as well supplied by geothermal energy from the deep aquifer, and part of the high-temperature heat is supplied by the exhaust gases of a natural gas engine. Regarding the electricity, part of its requirement is supplied by the natural gas engine, by the MSWI plant though its share is much reduced, and the last part by hydro-electricity. The natural gas engine and the deep aquifer are as well selected during the interseason periods (see Figures E.48 and E.49 in Annex E).

This example illustrates the validity of the methodology for adapting the system operating conditions to the seasonal variations, as well as the relevance of using a multi-period approach. The method could however be enhanced by including material and heat storage possibilities, which would allow for decoupling the production from the usage. An application would be the SNG production from wood during a period where the CHP production from this technology makes sense for the energy service requirement, and the storage of the produced SNG to distribute it uniformly over the year in transport.

An other application of the developed methodology is the identification of optimal pathways for resource and waste conversion, accounting as well for the limited availability of waste and of indigenous resources, and the services that can be produced from each resource or waste. This is illustrated by the example of configuration D2 and its biomass usage, which includes organic waste and woody biomass. The system operation per capita is displayed in Figure 5.8 for the first summer period – May, July and August. The pathways for biomass conversion are highlighted and their material and energy flows are displayed in color, while the other technologies and flows are shaded and their material and energy flows are displayed in grey. The biomethanation unit and the biogas engine are as well used in all other periods, while the wood-to-SNG conversion unit is not used during the second summer period – June – but during all the other ones (see Figures E.42 to E.45 in Annex E).

For environomic configurations, the optimal pathway for woody biomass is its conversion to combined electricity, heat and SNG, as it can be seen from Figure

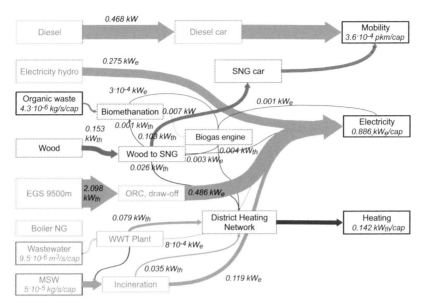

Figure 5.8 Example of optimal pathways for organic waste and woody biomass conversion for configuration D2 during period 1 (May, July & August) (adapted from Gerber et al. (2013)).

5.8. The SNG is then used for transport, and the woody biomass contributes therefore to supply the three services. The optimal environomic pathway for organic waste is the biomethanation to convert it to biogas, which is then used for cogeneration of heat and of electricity. Furthermore, the diagram highlights as well the limited potential for biomass. Indeed, the contribution to electricity and heat requirement of organic waste is relatively insignificant. Regarding woody biomass used to its full potential, which is the case for configuration D2, its contribution to the transport is of 14% for the whole year, its contribution to the district heating requirement between 18% and 5%, depending on the period, and its contribution to the electricity requirement is inferior to 1%. The relative reduction in yearly CO_2 emissions that can be attributed directly to the optimal biomass usage is around 5% when compared with the reference scenario.

By analyzing the points of the Pareto curve of Figure 5.4, it is possible to identify the pathways that are selected in some of the optimal configurations for biomass conversion, and the ones that are never selected and therefore suboptimal. This is illustrated by Figure 5.9, which shows, among all the potential pathways for woody biomass (Figure 5.9(a)) and for organic waste (Figure 5.9(b)), which ones are selected in some of the optimal configurations. The non-selected suboptimal pathways are shaded. Only the main energy and material flows are displayed in bright colors.

Regarding woody biomass, the only selected pathways in the optimal systems configurations are the conversion to SNG, electricity and heat for transport

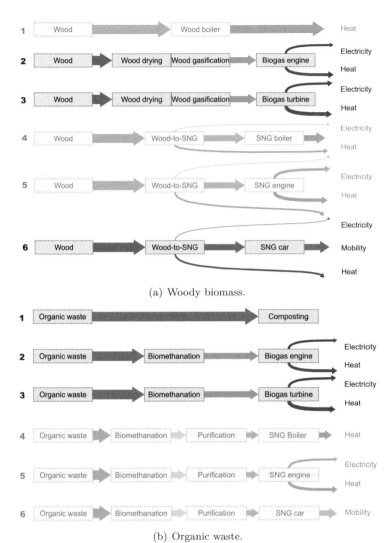

(a) Woody biomass.

(b) Organic waste.

Figure 5.9 Selected pathways in the optimal configuration of the Pareto curve of Figure 5.4 for biomass conversion: (a) woody biomass (b) organic waste. The suboptimal non-selected pathways are shaded (adapted from Gerber et al. (2013)).

applications (pathway 6 in Figure 5.9(a)) in most of the environomic configurations, and, in a few cases for clusters A, B and C, its gasification and use in a biogas engine or turbine for CHP (pathways 2 and 3 in Figure 5.9(a)). In general, the transport usage of the biomass is favored over the other ones, since other alternatives are available from the superstructure of technologies and resources for heat and electricity production, such as the MSWI plant or the geothermal technologies. The production of heat in a wood boiler (pathway 1 in Figure 5.9(a)), which is used in the current situation, or the production of SNG and its use in a boiler for heat production (pathway 4 in Figure 5.9(a)), or the use of SNG in an engine for CHP (pathway 5 in Figure5.9(a)), are thus not optimal pathways in the present case and are not selected by the optimizer.

For organic waste, the production of biogas through biomethanation and its purification to SNG quality (pathways 4, 5 and 6 in Figure 5.9(b)) are not selected due to the too high investment cost linked with a purification unit and the very limited potential of organic waste. Thus, the optimal pathways for organic waste treatment are either the composting (pathway 1 in Figure 5.9(b)), which is used in the current situation, or the biomethanation and the use of the produced biogas in an engine or a turbine for CHP (pathways 2 and 3 in Figure 5.9(b)).

These examples highlight the importance of accounting for the limited availability of indigenous resources and waste, and to select thereby the most efficient conversion pathway, as well as the best allocation between the different energy services that can be produced with a given resource or technology.

The developed methodology allows as well for selecting the optimal technologies for resource valorization. This is illustrated by the example of the technologies for geothermal energy conversion. Table 5.5 shows which ones of the embedded technologies in the superstructure are selected in some of the optimal configurations of the Pareto curve of Figure 5.4.

Table 5.5 Selection of technologies for geothermal energy conversion embedded in the superstructure for the optimal system configurations of the Pareto curve of Figure 5.4.

Pathway	Represented in Pareto
Deep aquifer 1450m, direct heat usage	yes
Deep aquifer 1450m, heat pumps	no
EGS 6000m, Kalina for CHP	yes
EGS 6500m, supercritical ORC for electricity production	yes
EGS 8000m, 2-flash for CHP	no
EGS 9500m, ORC with draw-off for CHP	yes
EGS 9500m, 1-flash & bottoming ORC for electricity production	yes

In the case of the deep aquifer, since the minimal return temperature of the district heating network is considered (38 to 45°C, see Table 5.3), and that the temperature of the geothermal water is 51°C, the geothermal heat is directly used to supply part of the low temperature heat requirement of the district heating network. The use of heat pumps for supplying higher temperature heat from the aquifer, which would increase the investment costs and the electricity consumption, is thus not considered in the optimal configurations. In the case of the EGS, the only technology not selected in the optimal configurations is the double-flash for CHP with an EGS at 8000m. Indeed, this solution is suboptimal compared to the other technologies for the conversion of geothermal heat from the EGS in terms of operating costs, due to the high maintenance costs linked with the use of a large double-flash system, and in terms of environmental impacts, due to the high fossil CO_2 emissions from the condensers of the double-flash system, as discussed previously in Chapter 4.

The methodology is therefore useful to identify the optimal technologies with respect to the defined objectives and to select the best ones that are adapted to the specific conditions of the case study, like the district heating network temperature.

A last important aspect illustrated by the application of the developed methodology to the case study is the use of a systems approach accounting for the interactions between the different supply chains. This allows for identifying the potential competitions or synergies between different resources and technologies.

An example of competition is the supplementary heat requirement for the district heating in winter, which can not be fully satisfied by the MSWI plant and the WWTP together. This supplementary heat requirement can either be supplied by the deep aquifer, or a technology using fossil resources like a natural gas engine or a boiler, or the wood-to-SNG conversion unit or the heat from the conversion cycle of the EGS if this one is used for CHP, all these pathways being competing.

This is illustrated by Figure 5.10, showing the system operation during the winter period – November to February – for two typical configurations using different technologies for supplying the supplementary heat requirement. The other technologies are shaded, and their material and energy flows are displayed in grey.

For the first configuration (see Figure 5.10(a)), the EGS is exclusively used for electricity production and the wood-to-SNG conversion unit is turned off. Moreover, the full potential of low-temperature heat from the WWTP is used. Other heat sources have therefore to be used to supply the requirement of the district heating, and these are the deep aquifer and the engine using fossil natural gas. For the second configuration (see Figure 5.10(b)), the EGS is used for CHP and the wood-to-SNG conversion unit, which produces as well heat, is turned on. These two heat sources and part of the heat available from the WWTP are sufficient to supply the district heating requirements not satisfied by the MSWI plant. Thus, the investment in the deep aquifer or in the natural gas engine are not necessary. Though this configuration allows for a full supply of the district heating network from waste or renewable sources, a part of the electricity requirement has to be supplied by the import of UCTE mix.

In all the studied typical optimal configurations, the deep aquifer is competing with the EGS configurations used for CHP and with the heat from the wood-to-SNG conversion unit. Indeed, it is never selected when an EGS configuration

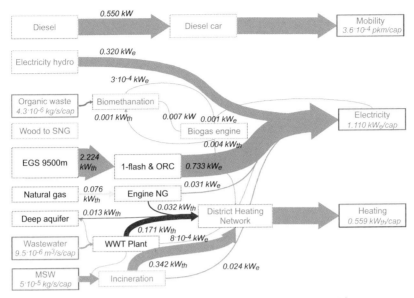

(a) Configuration E2 (EGS for electricity production only).

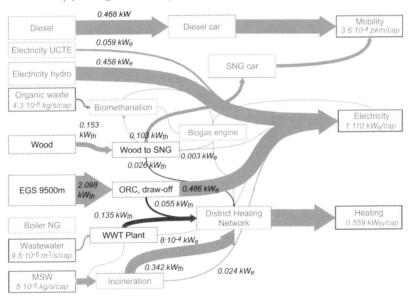

(b) Configuration D2 (EGS for heat and power production and wood-to-SNG conversion).

Figure 5.10 Heat production from different resources and technologies during period 5 (November-February) for two configurations: (a) Configuration E2 (b) Configuration D2 (adapted from Gerber et al. (2013)).

with CHP is selected, and even if it is selected, it is not used during the periods when the wood-to-SNG conversion unit is turned on (see Figures E.6 to E.55 of Annex E).

An example of synergy is between the MSWI plant and the WWTP during the first summer period – May, July and February. Indeed, using the heat produced by the WWTP reduces the supplementary requirement of the district heating network, which can be satisfied by using only a small part of the heat available from the MSWI plant (see Figure 5.7(a)). The remaining part of the heat from the waste incineration is thus available for an increased electricity production.

Such competitions or synergies could not have been highlighted without a systematic approach accounting for the interactions between the energy services requirements, the waste valorization possibilities and the pathways for the conversion of indigenous resources, as well as by using the layer approach described in section 2.3 to synthesize the supply chains for the different material and energy flows.

Optimal value of a CO_2 tax

In order to perform a final selection in the optimal configurations of the Pareto curve, the economic indicator of the payback period (Equation (5.4)) and the environmental indicator of the reduction in CO_2 emissions (Equation (5.5)) are used.

As displayed in Figure 5.11 below, if the raw results from the optimization are used, with a CO_2 tax almost equal to zero for all the optimal configurations, the economic optimum is the configuration A1 ($t_{pb} = 0.4$ yr), with almost no investment and only an adjustment of the system operating conditions, which is the worse one from an environmental point of view, since it increases slightly the CO_2 emissions ($\delta_I = -2\%$) due to the import of UCTE electricity mix.

In order to move the economic optimum to a configuration achieving a better environmental performance, the CO_2 tax can be adjusted and the payback period recalculated.

In the present case, the configuration A1 remains the economic optimum up to a CO_2 tax corresponding to 58 €/ton of CO_2-eq. As displayed in Figure 5.11, which shows as well the payback period for a CO_2 tax of 60 €/ton of CO_2-eq, the economic optimum becomes then the configuration E1 ($t_{pb} = 6.0$ yr), which has an environmental performance ($\delta_I = 36\%$) closer to the environmental optimum, the configuration D2 ($t_{pb} = 7.5$ yr, $\delta_I = 45\%$).

For a CO_2 tax corresponding to 60 €/ton of CO_2-eq, the environmental penalty to pay for choosing the economic optimum (E1) instead of the environmental optimum (D2) is a 20% relative decrease of the CO_2 reduction, while the economic penalty for choosing the environmental optimum instead of the economic one is a 25% increase in the payback period, which is a higher relative penalty. If one wanted to give a similar importance to the economic and environmental performance and share the penalty between them for not choosing one opti-

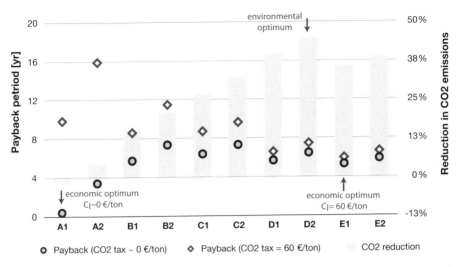

Figure 5.11 Reduction in CO_2 emissions and payback period of typical optimal configurations from the Pareto curve of Figure 5.4 (adapted from Gerber et al. (2013)).

mum or the other, the configuration D1, with the EGS for CHP at 9500m without wood-to-SNG conversion, would realize the best trade-off between the two criteria, with relative penalties of 10% and 12% for the economic and environmental performances, respectively.

Though the optimal value for the CO_2 tax identified by the multi-objective optimization is around zero, this analysis suggests however that such a tax is still necessary for selecting environomic system configurations rather than pure economic ones in the final design and planning of urban energy systems.

5.4 Conclusions on urban systems

In this chapter, it was presented how the developed methodology for the integration of LCA in the design of renewable energy systems has been applied to the design of an urban energy system, considering an extended action system and the identification of industrial ecology possibilities. The chosen application case study integrates the multiple energy services to be supplied with their seasonal variations, the waste to be treated, some of the potential indigenous resources that can be exploited – i.e. geothermal aquifers, EGS and woody biomass – and the different conversion technologies that can be used to treat the waste or convert the resources into useful energy services. Both detailed models with flowsheeting software and average technologies, based on the information of the LCI database, have been considered, and environomic criteria have been taken into account for the multi-objective optimization.

The results for the city taken as an application case study suggest that by selecting the adequate resources, technologies and adapting the system operating conditions, the life-cycle CO_2 emissions can be reduced down to 45% and the operating costs down to 59% when compared with the current situation. In addition, these results indicate that in the future perspective of replacing nuclear power in Switzerland or in any other country taking a similar orientation in terms of energy policy, EGS might be an economically competitive solution to avoid a large increase in the CO_2 emissions by a massive import of fossil-based electricity mix. Other more general outcomes with respect to the application of the methodology to the design of urban energy systems should as well be discussed.

A first major outcome is that the integration of an environmental objective in the optimization procedure in addition to the economic ones leads again to consider a panel of optimal environomic configurations that would not be considered otherwise in a pure economic optimization. This highlights the importance of fully integrating the environmental impact in the design procedure in order to identify solutions leading to an effective impact reduction: while the maximum impact reduction for pure economic solutions is of 39%, a higher impact reduction up to 45% is achieved with environomic solutions. Hence, resources and technologies allowing to reach a higher impact reduction that would not be selected otherwise in a pure economic optimization are represented in the optimal configurations. This is for example the case of wood usage as a renewable resource and its thermochemical conversion into Synthetic Natural Gas (SNG), electricity and heat.

Secondly, the case study illustrates the necessity of considering an extended action perimeter for complex systems, embedding the different services to be supplied, the waste to be treated, the available indigenous or imported resources and the potential conversion technologies as separate units in the superstructure. This allows for synthesizing the optimal supply chains for the valorization of waste and resources, by accounting for the seasonal variations, for the limited availability of waste and of indigenous resources, and for the possible competitions and synergies between the different supply chains.

In the present case, the operation of the municipal solid waste incineration plant and of the wastewater treatment plant, which are used for Combined Heat and Power production, can be adapted to the seasonal demand in district heating in order to minimize the consumption of other resources for heat supply. The optimal pathway for woody biomass conversion is a good example for the valorization of a limited indigenous resource: since other local and competitive alternatives are available for electricity and district heating production, such as the waste treatment facilities and the geothermal resources in our case, the best usage of this resource is its conversion to SNG and its use in transportation, for which no other low-carbon alternatives are available. It is clear that the best allocation of a resource depends on the different process alternatives considered in the superstructure. For example, Steubing et al. (Steubing

(2011); Steubing et al. (2012)) found different results with the best allocation of available wood in Europe being in boilers. Indeed, in comparison to this study, they did not consider in their list of options the combined production of heat with the SNG, the presence of alternative renewable energy resources and the advanced conversion systems such as heat pumps. This study highlights also that the decision-making is different at a local level considering the specific conditions and context, such as the city of La Chaux-de-Fonds. Such conditions are difficult to capture when regional or national levels are considered. This example of wood allocation suggests nevertheless the importance to account for the different competitors to supply a given service when evaluating different pathways for the valorization of a limited resource, and to study the adequacy between the resource availability and the demand in energy services.

Finally, it is shown that the methodology can be used to determine the optimal value of environmental taxes in order to favor the selection based on economic criteria of configurations that are closer to the environmental optimum. In the present case study, the two optimums are in conflict if no CO_2 tax is applied. Here, the minimal value of the CO_2 tax to switch the economic optimum closer to the environmental one is 60 €/ton of CO_2-eq.

Chapter 6

Conclusions

In this book, a systematic methodology for the integration of Life Cycle Assessment (LCA) in the conceptual design of renewable energy conversion technologies has been outlined, and combined with process design, process integration and multi-objective optimization techniques. The approach described is therefore inter-disciplinary by integrating together concepts coming both from the fields of process systems design and of environmental analysis. Indeed, the methodology allows to account for the environmental aspects at an early development stage, and to relate them to the economic and thermodynamic aspects. In this regard, it is well suited to identify preliminary designs that are promising for the development of emerging technologies in the field of renewable energies. Other applications are to orientate the decision-making linked with territorial planning of urban energy systems or with the identification of the best pathways for resource and waste valorization to be favored in future energy policies.

In the proposed methodology, the life cycle inventory elements are expressed as a function of the process operating conditions. This allows for obtaining indicators that reflect the environmental performance of the process design and configuration, by analogy with the thermo-economic indicators that are calculated in the same way. Therefore, the trade-offs between environmental impacts and economic or thermodynamic objectives can be calculated, and the environmental aspects can be fully integrated in the process design procedure. The extension of the decision perimeter to the supply chain synthesis and the identification of the industrial ecology possibilities are as well integrated in the approach. The different aspects of the developed methodology have been illustrated by three application case studies in the economic, political and geographic context of Switzerland. The first one concerned the optimal process design for the thermochemical conversion of wood to Synthetic Natural Gas (SNG) and electricity. The second one dealt with the optimal design of Enhanced Geothermal Systems (EGS) considering combined heat and power (CHP) production, in terms of targeted depths, conversion technologies and district heating design sizes to be favored. The last case study concerned the design of an urban energy conversion system considering the multiple energy services to be supplied, the waste to be treated, and the indigenous renewable resources that can potentially be used. The major outcomes and the perspectives opened by the development of the methodology and its application are summarized in the paragraphs below.

A first important outcome is that integrating the LCA in the process design procedure allows for obtaining indicators that reflect better the environmental performance of an industrial process with a commercial design. Furthermore, a proper process design and integration allows for an important impact reduction. This is well illustrated by the example of the thermochemical conversion of wood to SNG and electricity, for which a LCA based on a pilot-scale process has been compared with the proposed methodology integrating LCA in a thermo-economic model of a similar process accounting for process integration. Therefore, the method allows one to really calculate the environmental impact of a given process with a design close to the commercial one, rather than deducing this impact from a process design realized for another purpose.

A second contribution is the possibility to calculate the trade-offs between environmental, economic and thermodynamic objectives for identifying the promising process configurations. The future development of technologies or the decision-making can be thus oriented to account for the environmental aspects. An illustrative example is again provided by the thermochemical conversion of wood, which highlights a trade-off between the environmental impacts and the economic profitability linked with the process scale. Indeed, an increased scale reduces the costs due to the economies of scale, but increases at the same time the impact due to the process logistics and to the consumption of auxiliary materials. However, this impact of auxiliary materials can be reduced by switching to pressurized processes with a more compact design, which increases simultaneously the efficiency, and has a beneficial effect on both the costs and the impacts. Therefore, such process design options should be favored for the future development of the technology.

In the regard of the two outcomes discussed above, it appears that the mutual exchange of information between the field of LCA and process design is beneficial for both communities. From the point of view of the LCA, the combined method allows for working with realistic and consistent data, and for conducting a detailed study on the variations in environmental impacts for the different possible evolutions and future generation of a technology. From the point of view of the process design, considering the environmental impact at an early stage of the design allows the engineers to efficiently propose pertinent impact mitigation measures and modifications in the process design.

A third important outcome of the present work is that the full integration of LCA in the process design procedure leads to take different decisions than if only economic and thermodynamic criteria are considered. Indeed, the examples of EGS and of urban systems highlight that by integrating the LCA in the decision-making procedure, process configurations that would otherwise not be considered with a conventional thermo-economic analysis are kept in the final candidate process configurations. For the case of EGS, though at relatively shallow depths the thermo-economic and environmental performance indicators all favor the same configurations – an EGS around 5500m to 6000m with a district heating size between 20 and 35 MW_{th} using a Kalina cycle – for deep EGS between 8500m and 9500m the thermo-economic criteria and the environmental criteria are conflicting. While the payback period and the exergy efficiency tend to favor the single electricity production with a single-flash system and a bottoming organic Rankine cycle (ORC), the two considered impact assessment methods for LCA favor large CHP systems with a district heating size of about 60 MW_{th} and an ORC with an intermediate draw-off. Furthermore, when compared with each other at 9500m, the environmental benefits are higher for choosing the large CHP system over the single electricity production than the thermo-economic benefits for choosing the electricity production over a large CHP. In such a case, one could therefore argue that the environmental criteria outweigh the thermo-economic ones, and that large CHP systems should be favored for the exploitation of deep EGS. The case study on urban energy systems provides other examples of optimal environomic system configurations that would not have been considered with pure economic criteria. This is illustrated by the system configurations using the wood-to-SNG conversion technology, which increases the investment and operating costs, but decreases simultaneously the life-cycle CO_2 emissions. Moreover, though it is often argued that increasing the thermodynamic efficiency is equivalent to reducing the environmental impact, the examples of the thermochemical conversion of wood and of the EGS both indicate that the thermodynamic and the environmental optimums are not necessarily corresponding, although the process efficiency is nevertheless a key-point for impact reduction. Such results suggest thereby that the environmental criteria should be integrated as separate objectives in the process design procedure.

Eventually, the case study on urban systems illustrates the necessity of considering an extended action perimeter for complex systems, embedding the different services to be supplied, the waste to be treated, the available indigenous or imported resources and the potential conversion technologies as separate units that can be selected or not. This allows for synthesizing the optimal supply chains for the exploitation of waste and resources, by accounting for the seasonal variations, for the limited availability of waste and of indigenous resources, and for the possible competitions and synergies between the different supply chains. In the present case, the operation of the municipal solid waste incineration plant and of the wastewater treatment plant, which are used for CHP, can be adapted to the seasonal demand in district heating in order to minimize the consumption of other resources for heat supply. The selection of the woody biomass conversion technology is a good example for identifying the best usage of a limited indigenous resource. Since other local and competitive alternatives are available for electricity and district heating production, such as the waste treatment facilities and the geothermal resources, the best usage of this resource is its conversion to SNG and its use in transport, for which no other low-carbon alternatives are available. It is clear however that the best allocation of a resource depends on the different process alternatives and competing resources considered for a given application. The case study presented in this book highlights also that the decision-making is different at a local level considering the specific conditions and context, such as the city of La Chaux-de-Fonds. Indeed, this city features a municipal solid waste incineration plant of a certain capacity and technology, a given amount of available biomass, and geothermal resources with specific geological characteristics. Such conditions are difficult to capture when regional or national levels are considered. Another example is the EGS. For the case study specific to the EGS technology, the CHP systems with a large district heating using deep EGS are environmentally more favorable and ensure a better coverage for the conjoined demand in electricity and in district heating than single electricity production at the same depth. However, when this technology is evaluated in an urban systems context with competitors, the environmental advantage of the CHP system over single electricity production is reduced, since there is competition for the heat supply between the EGS, the waste treatment facilities and the other indigenous resources. These two examples pinpoint the importance to account for the different competitors to supply a given service when evaluating different pathways for the exploitation of a limited resource, and to perform the adequacy between the resource availability and the demand in energy services.

Furthermore, applying the developed methodology to the case studies has highlighted the importance of several aspects specifically linked with the LCA methodology. These aspects concern the functional unit definition, the choice of the life cycle impact assessment (LCIA) method and thus of the environmental objective function, the assumptions for the substitution of the produced energy services and the issues linked with the use of a LCI database. Indeed, the functional unit, valid both for the environmental and the economic performance indicators, has to be defined carefully depending on the type of problem to be solved. This can be illustrated by comparing the two case studies on the thermochemical conversion of wood and on the EGS. In both cases, though the first goal of these technologies is to produce a given energy service in priority – SNG for the former and electricity for the latter – the potential additional energy services that can be supplied by these processes – electricity for the former and heat for the latter – are however considered in the present case studies. Therefore, the function of such systems has to be defined with respect to the conversion of a given resource in multiple energy services, independently on which one of these services is considered *a priori* to be predominant. Moreover, the type of resource and its accessibility have to be accounted for in the functional unit definition. Though for the thermochemical wood conversion, the same resource is considered for any evaluated configuration, for the EGS, the resource is varying in quality – i.e. heat load and temperature level – depending on the targeted depth. Hence, the MJ of geothermal heat can not be taken as the functional unit like the MJ of wood.

A second aspect linked with the application of the LCA methodology is the choice of the LCIA method and therefore of the environmental objective function. Indeed, the case study on the thermochemical wood conversion shows that depending on which impact assessment method is selected as the objective function, the optimal process configurations favor either the production of SNG or of electricity, since the two services are weighted differently in each impact assessment method. This leads to discuss the third issue specifically linked with the application of the LCA methodology, concerning the substitution of the energy services, which is necessary to account for the process efficiency when evaluating a technology. Indeed, for the two case studies considering substitution, the thermochemical wood conversion and the EGS, the balance between the harmful environmental impacts and the beneficial ones from the substitution highlights the importance of the latter and, in the case of multiple energy services, that generally one service is favored over the other. The optimal ratio between the produced energy services depends on the process operating conditions and on the selected technology, on the choice of the impact assessment method, but as well on the assumption made for its substitution. The case of electricity is a good example. Indeed, the assumption for substitution was changed from a mix of nuclear power and import of fossil-based power for the thermochemical wood conversion to the production from a natural gas combined cycle for the EGS, due to the decision of the Swiss Federal Council to progressively abandon nuclear power that happened in between. This choice has important consequences on the optimal configurations, since the nuclear waste are more or less strongly weighted depending on the impact assessment method. Therefore, the electricity generation is favored or not over the other potential energy services produced by the technology. The replacement of nuclear power has been further addressed in the case study of urban systems. In this context, the results indicate that EGS might be competitive to avoid a large increase in the CO_2 emissions and in the operating costs linked with electricity consumption, due to the import of fossil-based electricity mix. This illustrates the potential consequences of a political choice on engineering decisions.

The last issue linked with the LCA methodology concerns the use of LCI databases, and more specifically of ecoinvent®. Regarding the impact scaling laws, there is currently a lack of datasets for process equipment, and when these are available, it is generally only for one or two different sizes. Though this is sufficient to apply the methodology for establishing scaling laws, more datasets would allow for increasing the confidence level in such laws. A second issue linked with the use of LCI databases in the present context of renewable energy systems is that an important part of the impact is often due to off-site emissions, like this is the case of the auxiliary materials for the thermochemical wood conversion. Though this issue is partially solved by the extension of the decision system to the supply chain synthesis, there is a cut-off level where no control is possible anymore and where the average market technology has to be accepted with its uncertainty and without any possibility of reducing the impact anymore. Finally, although the developed methodology is potentially applicable to any other economic, geographical and political context than Switzerland, finding reliable geo-localized LCI data might be tricky depending on the location, due to the current limitations of LCI databases, ecoinvent® containing mostly data specific to the Swiss or to the European context.

In addition to these important outcomes, the present work opens as well perspectives in the field of energy systems design and evaluation, and the methodology should be extended in the future with other important aspects that have not been detailed in this book. First of all, although the developed methodology has been illustrated with three specific application case studies, it is potentially applicable to any other renewable energy technology, though depending on the location, finding reliable LCI data can be difficult. The localization of LCI data is indeed a key aspect of any LCA application, and many data used in the present work depend on it, such as the logistics for lignocellulosic biomass, the geothermal gradient and the geological profile for geothermal resources, or the existence of waste treatment facilities for urban systems. Though some of these data, like the demand profiles in district heating, are already based on Geographic Information Systems (GIS), a complete and dynamic integration of the computational framework with GIS would allow for extending the applications of the methodology to the selection of optimal locations for the technologies in adequacy with the resource availability, the logistics and with the local demand in energy services. For the application to the design of urban energy systems, this would allow for adding location constraints and to calculate in a systematic manner the impacts and the costs linked with the logistics. Therefore, the methodology could be applied to regions or entire countries. Another important aspect that should be included in the future is to extend the methodology to consider energy and mass storage. Indeed, decoupling the production of a service from its use might allow to improve the environomic performances of a system or of a technology. The uncertainties linked with the use of LCI data is another important topic that should be included in a future extension of the methodology. Two types of data are concerned: the average data of background processes from ecoinvent®, including already some uncertainty information that could be used, and the data specifically linked with the system operation or the technology, which are currently not always available. Finally, in addition to urban systems, the approach for performing supply chain synthesis and identifying industrial ecology possibilities could as well be suited to design eco-industrial parks and to identify the potential industrial symbioses in existing industrial areas.

ecoinvent® equivalences

A.1 Thermochemical wood conversion

Table A.1 ecoinvent® equivalences of the material and energy flows included in the LCA model for biomass production and SNG conversion process.

LCI element & Unit	ecoinvent® equivalence	Reference
Hard wood chips, in m^3	wood chips, hardwood, u=80%, at forest, RER	Werner et al. (2003)
Soft wood chips, in m^3	wood chips, softwood, u=140%, at forest, RER	Werner et al. (2003)
Transport from forest to SNG plant, in tkm	transport, lorry, 20-28t, fleet average, CH	Dones and Bauer (2007)
Empty transport from SNG plant to forest, in km	operation, lorry, 20-28t, empty, fleet average, CH	Dones and Bauer (2007)
Water for process, in kg	tap water, at user, CH	Althaus et al. (2007b)
Olivine, in kg	basalt, at mine, RER	Althaus et al. (2007a)
Charcoal, in kg	charcoal, at plant, GLO	Werner et al. (2003)
Calcium carbonate, in kg	limestone, milled, loose, at plant, CH	Kellenberger et al. (2007)
Starting oil, in kg	light fuel oil, at regional storage, CH	Dones et al. (2007)
Oxygen (cryogenic), in kg	oxygen, liquid, at plant, RER	Althaus et al. (2007b)
Solid waste disposal, in kg	disposal, inert material, 0% water, to sanitary landfill, CH	Doka (2007)
Zinc oxide catalyst, in kg	zinc oxide, at plant, RER	citeecoinvrep18
Rape methyl ester, in kg	rape methyl ester, at esterification plant, CH	Jungbluth et al. (2007)
Gypsum disposal, in kg	disposal, gypsum, 19.4% water, to sanitary landfill, CH	Doka (2007)
Zinc oxide disposal, in kg	disposal, inert material, 0% water, to sanitary landfill, CH	Doka (2007)
Nickel catalyst, in kg	nickel, 99.5%, at plant, GLO	Althaus et al. (2007a)
Alumin. oxide catalyst, in kg	aluminium oxide, at plant, RER	Althaus et al. (2007a)
Nickel disposal, in kg	disposal, inert material, 0% water, to sanitary landfill, CH	Doka (2007)
Aluminum oxide disposal, in kg	disposal, inert material, 0% water, to sanitary landfill, CH	Doka (2007)
Avoided fossil NG extraction, in Nm^3	natural gas, at long-distance pipeline, CH	Dones et al. (2007)
Consum./avoid. elec., in kWh	electricity mix, CH	Dones et al. (2007)
Transport of auxiliary materials and waste, in tkm	transport, lorry, 20-28t, fleet average, CH	Dones and Bauer (2007)

A.2   Enhanced Geothermal Systems

Table A.2 ecoinvent® equivalences of the material and energy flows included in the LCA model for EGS.

Name of LCI element & Unit	ecoinvent® equivalence	Reference
Diesel for site preparation, drilling rig drive, drilling mud and reservoir enhancement, in MJ	diesel, burned in building machine, GLO	Kellenberger et al. (2007)
Cement for site preparation, cementation and well dismantling, in kg	cement, unspecified, at plant, CH	Kellenberger et al. (2007)
Bentonite for drilling mud and cementation, in kg	bentonite, at processing, DE	Kellenberger et al. (2007)
Starch for drilling mud, in kg	modified starch, at plant, RER	Althaus et al. (2007c)
Chalk for drilling mud, in kg	limestone, milled, loose, at plant, CH	Kellenberger et al. (2007)
Water for drilling mud and cementation, in kg	water, decarbonised, at plant, RER	Dones et al. (2007)
Calcium carbonate for drilling mud, in kg	limestone, milled, loose, at plant, CH	Kellenberger et al. (2007)
Cuttings disposal, in kg	disposal, inert material, 0% water, to sanitary landfill, CH	Doka (2007)
High-alloyed steel for casing, in kg	chromium steel 18/8, at plant, RER	Althaus et al. (2007a)
Low-alloyed steel for casing, in kg	steel, low-alloyed, at plant, RER	Althaus et al. (2007a)
Portland cement for cementation, in kg	portland calcareous cement, at plant, CH	Kellenberger et al. (2007)
Silica sand for cementation, in kg	silica sand, at plant, DE	Kellenberger et al. (2007)
Water (demineralized) for reservoir enhancement and make-up for EGS, in kg	water, deionised, at plant, CH	Althaus et al. (2007b)
Hydrochloric acid for reservoir enhancement, in kg	hydrocholoric acid, 30% in H2O, at plant, RER	Althaus et al. (2007b)
Hydrochloric acid for reservoir enhancement, in kg	hydrocholoric acid, 30% in H2O, at plant, RER	Althaus et al. (2007b)
Transport by lorry, in tkm	transport, lorry, 20-28t, fleet average, CH	Dones and Bauer (2007)
ORC working fluid (n-butane, iso-butane, cyclo-butane), in kg	propane/butane, at refinery, CH	Dones et al. (2007)
ORC working fluid (n-pentane, iso-pentane), in kg	pentane, at plant, RER	Hischier (2007)
ORC working fluid (benzene), in kg	benzene, at plant, RER	Hischier (2007)
ORC working fluid (toluene), in kg	toluene, liquid, at plant, RER	Althaus et al. (2007b)
ORC working fluid (R134a), in kg	refrigerant R134a, at plant, RER	Dones et al. (2007)
Kalina working fluid (ammonia), in kg	ammonia, liquid, at regional storehouse, CH	Althaus et al. (2007b)
Kalina working fluid (water), in kg	water, deionised, at plant, CH	Althaus et al. (2007b)
Avoided electricity from NGCC, in kWh	natural gas, burned in combined cycle plant, best technology, RER	Dones et al. (2007)
Avoided district heating from natural gas condensing boiler, in MJ	natural gas, burned in boiler condensing modulating > 100kW, RER	Dones et al. (2007)
Gravel for well dismantling, in kg	gravel, unspecified, at mine, CH	Kellenberger et al. (2007)

A.3   Urban systems

Table A.3 ecoinvent® equivalences or references for the LCI models of the resources included in the superstructure of the case study for urban systems.

Model of the superstructure	ecoinvent® equivalence or parameters used for modeling	Reference
Woody biomass	wood chips, mixed, u=120%, at forest, RER, in m^3	Werner et al. (2003)
Deep aquifer	modeled as a function of depth $z = 1450m$ and of the number of wells $n_w = 2$	Gerber and Maréchal (2012c)
Hot Dry Rock	modeled as a function of depth associated with the configuration $z1 = 6000m$, $z2 = 6500m$, $z3 = 8000m$, $z4, z5 = 9500m$	Gerber and Maréchal (2012c)
Hydro-electricity	electricity, hydropower, at power plant, CH, in kWh$_e$	Dones et al. (2007)
Natural gas	natural gas, high pressure, at consumer, CH, in MJ	Dones et al. (2007)
Light fuel oil	light fuel oil, at regional storage, CH, in kg	Dones et al. (2007)
Petrol	petrol, unleaded, at regional storage, CH, in kg	Dones et al. (2007)
Diesel	diesel, at regional storage, CH, in kg	Dones et al. (2007)
Electricity UCTE mix	electricity, production mix UCTE, in kWh$_e$	Dones et al. (2007)

Table A.4 ecoinvent® equivalences or references for the LCI model of biomethanation included in the superstructure of the case study for urban systems.

Model of the superstructure	ecoinvent® equivalence or parameters used for modeling	Reference
Biomethanation	biogas, from biowaste, at storage, CH, in Nm^3	Jungbluth et al. (2007)

Table A.5 ecoinvent® equivalences or references for the LCI models of cars included in the superstructure of the case study for urban systems.

Model of the superstructure	ecoinvent® equivalence or parameters used for modeling	Reference
Diesel car	operation, passenger car, diesel, fleet average 2010, CH, in km	Dones and Bauer (2007)
Petrol car	operation, passenger car, petrol, fleet average 2010, CH, in km	Dones and Bauer (2007)
Natural gas car	operation, passenger car, natural gas, CH, in km	Dones and Bauer (2007)
SNG car	operation, passenger car, methane, 96 vol-%, from biogas, CH, in km	Dones and Bauer (2007)

Table A.6 ecoinvent® equivalences or references for the LCI models of boilers included in the superstructure of the case study for urban systems.

Model of the superstructure	LCI element	ecoinvent® equivalence or parameters used for modeling	Reference
Biomass boiler	biogenic CO_2 emissions	calculated by model	Fazlollahi and Maréchal (2013)
Biomass boiler	other emissions	wood chips, from forest, mixed, burned in furnace 50 kW, CH, in MJ_{th}	Dones et al. (2007)
Biomass boiler	boiler	calculated by model, with corresponding scaling law	Gerber et al. (2011a)
Light fuel oil boiler	fossil CO_2 emissions	calculated by model	Fazlollahi and Maréchal (2013)
Light fuel oil boiler	other emissions	light fuel oil, burned in boiler 10kW, condensing, non-modulating, RER, in MJ_{th}	Dones et al. (2007)
Light fuel oil boiler	boiler	calculated by model, with corresponding scaling law	Gerber et al. (2011a)
Natural gas boiler	fossil CO_2 emissions	calculated by model	Fazlollahi and Maréchal (2013)
Natural gas boiler	other emissions	natural gas, burned in boiler condensing modulating <100kW, RER, in MJ_{th}	Dones et al. (2007)
Natural gas boiler	boiler	calculated by model, with corresponding scaling law	Gerber et al. (2011a)
SNG boiler	biogenic CO_2 emissions	calculated by model	Fazlollahi and Maréchal (2013)
SNG boiler	other emissions	natural gas, burned in boiler condensing modulating <100kW, RER, in MJ_{th}	Dones et al. (2007)
SNG boiler	boiler	calculated by model, with corresponding scaling law	Gerber et al. (2011a)

Table A.7 ecoinvent® equivalences or references for the LCI models of conversion cycles for geothermal heat included in the superstructure of the case study for urban systems.

Model of the superstructure	ecoinvent® equivalence or parameters used for modeling	Reference
Kalina cycle	calculated by model	Gerber and Maréchal (2012c)
ORC-s, using R134a	calculated by model	Gerber and Maréchal (2012c)
double-flash system	calculated by model	Gerber and Maréchal (2012c)
single-flash system & ORC, using iso-butane	calculated by model	Gerber and Maréchal (2012c)
ORC-d, using cyclo-butane	calculated by model	Gerber and Maréchal (2012c)

Table A.8 ecoinvent® equivalences or references for the LCI models of biomass dryers included in the superstructure of the case study for urban systems.

Model of the superstructure	ecoinvent® equivalence or parameters used for modeling	Reference
Air dryer	calculated by model	Gerber et al. (2011a)
Steam dryer	calculated by model	Gerber et al. (2011a)

Table A.9 ecoinvent® equivalences or references for the LCI models of biomass gasifiers included in the superstructure of the case study for urban systems.

Model of the superstructure	ecoinvent® equivalence or parameters used for modeling	Reference
Air gasifier for engine	calculated by model	Gerber et al. (2011a)
Air gasifier for turbine	calculated by model	Gerber et al. (2011a)
Steam gasifier for engine	calculated by model	Gerber et al. (2011a)
Steam gasifier for turbine	calculated by model	Gerber et al. (2011a)

Table A.10 ecoinvent® equivalences or references for the LCI models of engines included in the superstructure of the case study for urban systems.

Model of the superstructure	LCI element	ecoinvent® equivalence or parameters used for modeling	Reference
Biogas engine	emissions	biogas, burned in cogen with gas engine, CH, in MJ_{th}	Jungbluth et al. (2007)
Biogas engine	engine	calculated by model, with corresponding scaling law	Gerber et al. (2011a)
Natural gas engine	emissions	natural gas, burned in cogen 50 kWe lean burn, CH, in MJ_{th}	Dones et al. (2007)
Natural gas engine	engine	calculated by model, with corresponding scaling law	Gerber et al. (2011a)
SNG engine	biogenic CO_2, CO and CH4 emissions	quantities based on the model using natural gas, but emissions changed to biogenic	Dones et al. (2007)
SNG engine	other emissions	natural gas, burned in cogen 50 kWe lean burn, CH, in MJ_{th}	Dones et al. (2007)
SNG engine	engine	calculated by model, with corresponding scaling law	Gerber et al. (2011a)

Table A.11 ecoinvent® equivalences or references for the LCI models of heat pumps included in the superstructure of the case study for urban systems.

Model of the superstructure	LCI element	ecoinvent® equivalence or parameters used for modeling	Reference
Heat pump	initial working fluid, emissions, maintenance, loss at end-of-life, and associated transports	heat, at air-water heat-pump 10 kW, CH, in MJ_{th}	Dones et al. (2007)
Heat pump	compressor and heat exchangers	calculated by model, with corresponding scaling law	Gerber et al. (2011a)

Table A.12 ecoinvent® equivalences or references for the LCI model of MSWI plant included in the superstructure of the case study for urban systems.

Model of the superstructure	ecoinvent® equivalence or parameters used for modeling	Reference
MSWI plant	disposal, municipal solid waste, 22.9% water, to municipal incineration, CH, in kg	Doka (2007)

Table A.13 ecoinvent® equivalences or references for the LCI model of purification and pressurization of biogas to SNG quality included in the superstructure of the case study for urban systems.

Model of the superstructure	ecoinvent® equivalence or parameters used for modeling	Reference
Purification	methane, 96 vol-%, from biogas, at purification, CH, in Nm^3	Jungbluth et al. (2007)
Pressurization	methane, 96 vol-%, from biogas, high pressure, at consumer, CH, in MJ	Jungbluth et al. (2007)

Table A.14 ecoinvent® equivalences or references for the LCI model of combined SNG, electricity and heat production from woody biomass included in the superstructure of the case study for urban systems.

Model of the superstructure	ecoinvent® equivalence or parameters used for modeling	Reference
SNG production	calculated by model	Gerber et al. (2011a)

Table A.15 ecoinvent® equivalences or references for the LCI model of biogas turbine included in the superstructure of the case study for urban systems.

Model of the superstructure	LCI element	ecoinvent® equivalence or parameters used for modeling	Reference
Biogas turbine	emissions	biogas, burned in micro gas turbine 100kWe, CH, in MJ_{th}	Primas (2007)
Biogas turbine	turbine	calculated by model, with corresponding scaling law	Gerber et al. (2011a)

Table A.16 ecoinvent® equivalences or references for the LCI model of WWTP included in the superstructure of the case study for urban systems.

Model of the superstructure	ecoinvent® equivalence or parameters used for modeling	Reference
WWTP	treatment, sewage, to wastewater treatment, class 3, CH, in m^3	Doka (2007)

Configurations for thermochemical wood conversion

Table B.1 Characteristics of the configuration using FICFB at 5 MW$_{th}$, with Ecoscarcity06 as the environmental objective (configuration 1 on Figure 3.5).

Drying technology	air
Biomass pretreatment	none
Gasification technology	atm. indirect steam-blown (FICFB)
Gas cleaning technology	cold
Process scale, in MW$_{th}$	5
Dryer temperature, in °K	513
Wood humidity after gasifier inlet	0.3
Gasification pressure, in bar	1.15
Gasification temperature, in °K	1123
Methanation pressure, in bar	5.26
Methanation inlet temperature, in °K	668
Methanation outlet temperature, in °K	601
SNG production, in MW	3.3
Electricity production, in MW$_e$	0.32
Energy efficiency, in SNG-eq	0.71
Biomass profitability, in €/MWh	1.04
Ecoscarcity06 impact, in UBP/MJ	−10.4

Table B.2 Characteristics of the configuration using FICFB at 200 MW$_{th}$, with Ecoscarcity06 as the environmental objective (configuration 2 on Figure 3.5).

Drying technology	air
Biomass pretreatment	none
Gasification technology	atm. indirect steam-blown (FICFB)
Gas cleaning technology	cold
Process scale, in MW$_{th}$	200
Dryer temperature, in °K	513
Wood humidity after gasifier inlet	0.2
Gasification pressure, in bar	1.15
Gasification temperature, in °K	1123
Methanation pressure, in bar	4.95
Methanation inlet temperature, in °K	670
Methanation outlet temperature, in °K	592
SNG production, in MW	131
Electricity production, in MW$_e$	7.9
Energy efficiency, in SNG-eq	0.71
Biomass profitability, in €/MWh	28.1
Ecoscarcity06 impact, in UBP/MJ	−7.69

Table B.3 Characteristics of the configuration using FICFB at 60 MW$_{th}$, with Ecoindicator99-(h,a) as the environmental objective (configuration 3 on Figure 3.5).

Drying technology	air
Biomass pretreatment	none
Gasification technology	atm. indirect steam-blown (FICFB)
Gas cleaning technology	cold
Process scale, in MW$_{th}$	60
Dryer temperature, in °K	512
Wood humidity after gasifier inlet	0.1
Gasification pressure, in bar	1.15
Gasification temperature, in °K	1123
Methanation pressure, in bar	8.65
Methanation inlet temperature, in °K	671
Methanation outlet temperature, in °K	576
SNG production, in MW	41.7
Electricity production, in MW$_e$	0.15
Energy efficiency, in SNG-eq	0.69
Biomass profitability, in €/MWh	23.9
Ecoindicator99-(h,a) impact, in pts/MJ	$−1.35 \cdot 10^{-3}$

Table B.4 Characteristics of the configuration using pFICFB at 60 MW$_{th}$, with Ecoindicator99-(h,a) as the environmental objective (configuration 4 on Figure 3.5).

Drying technology	air
Biomass pretreatment	none
Gasification technology	pres. indirect steam-blown (pFICFB)
Gas cleaning technology	cold
Process scale, in MW$_{th}$	60
Dryer temperature, in °K	513
Wood humidity after gasifier inlet	0.1
Gasification pressure, in bar	11.75
Gasification temperature, in °K	1123
Methanation pressure, in bar	
Methanation inlet temperature, in °K	671
Methanation outlet temperature, in °K	579
SNG production, in MW	41.1
Electricity production, in MW$_e$	-0.84
Energy efficiency, in SNG-eq	0.66
Biomass profitability, in €/MWh	26.1
Ecoindicator99-(h,a) impact, in pts/MJ	$-1.82 \cdot 10^{-3}$

Table B.5 Characteristics of the configuration using CFB, hcl at 60 MW$_{th}$, with Ecoindicator99-(h,a) as the environmental objective (configuration 5 on Figure 3.5).

Drying technology	steam
Biomass pretreatment	none
Gasification technology	pres. direct oxygen-blown (CFB)
Gas cleaning technology	hot
Process scale, in MW$_{th}$	60
Dryer temperature, in °K	453
Wood humidity after gasifier inlet	0.1
Gasification pressure, in bar	30.95
Gasification temperature, in °K	1073
Methanation pressure, in bar	30.15
Methanation inlet temperature, in °K	673
Methanation outlet temperature, in °K	575
SNG production, in MW	45.4
Electricity production, in MW$_e$	0.66
Energy efficiency, in SNG-eq	0.78
Biomass profitability, in €/MWh	41.5
Ecoindicator99-(h,a) impact, in pts/MJ	$-2.06 \cdot 10^{-3}$

Table B.6 Characteristics of the configuration using FICFB at 12 MW$_{th}$, with Ecoscarcity06 as the environmental objective (configuration 6 on Figure 3.5).

Drying technology	air
Biomass pretreatment	none
Gasification technology	atm. indirect steam-blown (FICFB)
Gas cleaning technology	cold
Process scale, in MW$_{th}$	12
Dryer temperature, in °K	513
Wood humidity after gasifier inlet	0.3
Gasification pressure, in bar	1.15
Gasification temperature, in °K	1173
Methanation pressure, in bar	5.17
Methanation inlet temperature, in °K	672
Methanation outlet temperature, in °K	600
SNG production, in MW	7.9
Electricity production, in MW$_e$	0.76
Energy efficiency, in SNG-eq	0.71
Biomass profitability, in €/MWh	13.7
Ecoscarcity06 impact, in UBP/MJ	−9.94

Table B.7 Characteristics of the configuration using FICFB, tor at 12 MW$_{th}$, with Ecoscarcity06 as the environmental objective (configuration 7 on Figure 3.5).

Drying technology	air
Biomass pretreatment	torrefaction
Gasification technology	atm. indirect steam-blown (FICFB)
Gas cleaning technology	cold
Process scale, in MW$_{th}$	12
Dryer temperature, in °K	513
Wood humidity after gasifier inlet	0.29
Gasification pressure, in bar	1.15
Gasification temperature, in °K	1173
Methanation pressure, in bar	4.81
Methanation inlet temperature, in °K	668
Methanation outlet temperature, in °K	612
SNG production, in MW	8.4
Electricity production, in MW$_e$	0.44
Energy efficiency, in SNG-eq	0.70
Biomass profitability, in €/MWh	11.9
Ecoscarcity06 impact, in UBP/MJ	−9.29

Table B.8 Characteristics of the configuration using FICFB at 12 MW$_{th}$, with Ecoindicator99-(h,a) as the environmental objective (configuration 8 on Figure 3.5).

Drying technology	air
Biomass pretreatment	none
Gasification technology	atm. indirect steam-blown (FICFB)
Gas cleaning technology	cold
Process scale, in MW$_{th}$	12
Dryer temperature, in °K	513
Wood humidity after gasifier inlet	0.1
Gasification pressure, in bar	1.15
Gasification temperature, in °K	1173
Methanation pressure, in bar	7.88
Methanation inlet temperature, in °K	673
Methanation outlet temperature, in °K	576
SNG production, in MW	8.6
Electricity production, in MW$_e$	0.02
Energy efficiency, in SNG-eq	0.69
Biomass profitability, in €/MWh	13.8
Ecoindicator99-(h,a) impact, in pts/MJ	$-1.41 \cdot 10^{-3}$

Table B.9 Characteristics of the configuration using FICFB, tor at 12 MW$_{th}$, with Ecoindicator99-(h,a) as the environmental objective (configuration 9 on Figure 3.5).

Drying technology	air
Biomass pretreatment	torrefaction
Gasification technology	atm. indirect steam-blown (FICFB)
Gas cleaning technology	cold
Process scale, in MW$_{th}$	12
Dryer temperature, in °K	512
Wood humidity after gasifier inlet	0.3
Gasification pressure, in bar	1.15
Gasification temperature, in °K	1173
Methanation pressure, in bar	12.61
Methanation inlet temperature, in °K	667
Methanation outlet temperature, in °K	591
SNG production, in MW	8.7
Electricity production, in MW$_e$	-0.18
Energy efficiency, in SNG-eq	0.68
Biomass profitability, in €/MWh	13.4
Ecoindicator99-(h,a) impact, in pts/MJ	$-1.54 \cdot 10^{-3}$

Configurations for EGS

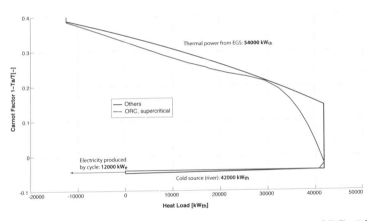

Figure C.1 Integrated exergy composite curve of the supercritical ORC with R134a for single electricity production at 6500m (config. 1 on Figure 4.13).

Table C.1 Operating conditions of the supercritical ORC with R134a for single electricity production at 6500m (config. 1 on Figure 4.13), (adapted from Gerber and Maréchal (2012c)).

Temperature at extraction well, in °C	219
Temperature at reinjection well, in °C	80
Thermal power available from EGS, in MW$_{th}$	54.0
Exergy available from EGS, in MW	17.5
Higher pressure of supercritical ORC, in bar	55
Superheating temperature of supercritical ORC, in °C	209
Electrical power produced by cycle, in MW$_e$	12.0
Parasitic losses, in MW$_e$	1.1
Net electrical power produced, in MW$_e$	10.9
Energy/Electrical efficiency, in %	20.2
Exergy efficiency, in %	62.3

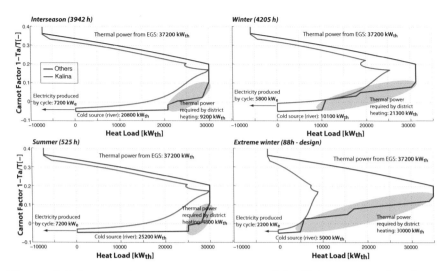

Figure C.2 Integrated exergy composite curve of the Kalina cycle with a district heating design size of 30 MW at 6000m (config. 2 on Figure 4.13).

Table C.2 Operating conditions of the Kalina cycle with a district heating design size of 30 MW at 6000m (config. 2 on Figure 4.13), (adapted from Gerber and Maréchal (2012c)).

	Interseason	Winter	Summer	Design
Operating time, in h	3942	4205	525	88
Temperature at extraction, in °C	201	201	201	201
Temperature at reinjection, in °C	106	106	106	106
Thermal power available from EGS, in MW_{th}	37.2	37.2	37.2	37.2
Exergy available from EGS, in MW	12.4	12.4	12.4	12.4
Higher pressure of Kalina cycle, in bar	36	36	36	36
Lower pressure of Kalina cycle, in bar	6	7.5	6	6
Ammonia concentration in working fluid of Kalina cycle	0.805	0.805	0.805	0.805
District heating demand, in MW_{th}	9.2	21.3	4.8	30
Electrical power produced by cycle, in MW_e	7.2	5.8	7.2	2.2
Parasitic losses, in MW_e	1.1	1.1	1.1	1.1
Net electrical power produced, in MW_e	6.1	4.7	6.1	1.1
Energy efficiency, in %	41.2	70.0	29.4	83.8
Electrical efficiency, in %	16.5	12.7	16.5	3.0
Exergy efficiency, in %	54.8	54.1	51.9	38.1

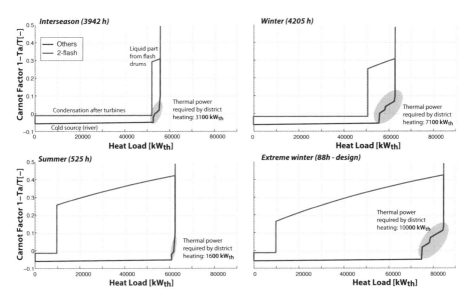

Figure C.3 Integrated exergy composite curve of the double-flash with a district heating design size of 10 MW at 8000m (config. 3 on Figure 4.13).

Table C.3 Operating conditions of the double-flash system with a district heating design size of 10 MW at 8000m (config. 3 on Figure 4.13), (adapted from Gerber and Maréchal (2012c)).

	Interseason	Winter	Summer	Design
Operating time, in h	3942	4205	525	88
Temperature at extraction well, in °C	272	272	272	272
Temperature at reinjection well, in °C	116	102	129	85
Thermal power available from EGS, in MW_{th}	63.3	68.3	58.1	74.9
Exergy available from EGS, in MW	24.5	25.8	23.1	27.2
Flashing temperature of 1st flash drum, in °C	215	217	258	268
Flashing temperature of 2nd flash drum, in °C	165	169	256	256
District heating demand, in MW_{th}	3.1	7.1	1.6	10
Electrical power produced by cycle, in MW_e	16.6	16.3	3.8	3.8
Parasitic losses, in MW_e	1.1	1.1	1.1	1.1
Net electrical power produced, in MW_e	15.5	15.2	2.7	2.7
Energy efficiency, in %	29.4	32.7	7.4	17.0
Electrical efficiency, in %	24.5	22.3	4.7	3.7
Exergy efficiency, in %	64.0	61.5	12.2	14.5

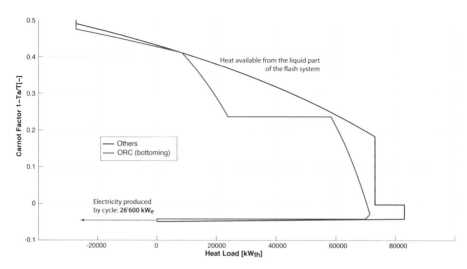

Figure C.4 Integrated exergy composite curve of the single-flash with a bottoming ORC with iso-butane for single electricity production at 9500m (config. 4 on Figure 4.13).

Table C.4 Operating conditions of the 1-flash with a bottoming ORC with iso-butane for single electricity production at 9500m (config. 4 on Figure 4.13), (adapted from Gerber and Maréchal (2012c)).

Temperature at extraction well, in °C	324
Temperature at reinjection well, in °C	84
Thermal power available from EGS, in MW$_{th}$	101.5
Exergy available from EGS, in MW	40.0
Flashing temperature of 1st flash drum, in °C	313
Evaporation temperature of ORC, in °C	114
Superheating temperature of ORC, in °C	294
Electrical power produced by flash, in MW$_e$	4.0
Electrical power produced by ORC, in MW$_e$	26.6
Parasitic losses, in MW$_e$	1.1
Net electrical power produced, in MW$_e$	29.5
Energy/Electrical efficiency, in %	29.1
Exergy efficiency, in %	73.7

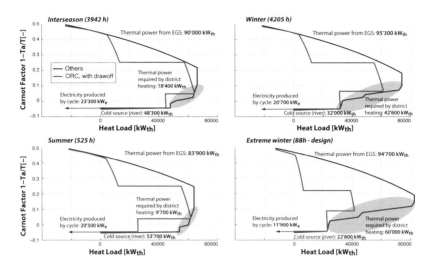

Figure C.5 Integrated exergy composite curve of the ORC with an intermediate draw-off using cyclo-butane, with a district heating design size of 60 MW at 9500m (config. 5 on Figure 4.13).

Table C.5 Operating conditions of the ORC with an intermediate draw-off using cyclo-butane, with a district heating design size of 60 MW at 9500m (config. 5 on Figure 4.13), (adapted from Gerber and Maréchal (2012c)).

	Interseason	Winter	Summer	Design
Operating time, in h	3942	4205	525	88
Temperature at extraction, in °C	324	324	324	324
Temperature at reinjection, in °C	114	100	130	101
Thermal power available from EGS, in MW_{th}	90.0	95.3	83.9	94.7
Exergy available from EGS, in MW	37.4	38.7	35.6	38.6
Evaporation temp. of ORC, in °C	125	127	125	109
Superheating temp. of ORC, in °C	308	311	313	309
Condensation temp. of draw-off in ORC, in °C	41	49	40	56
Splitting fraction for intermediate draw-off of ORC, in %	32	51	60	48
District heating demand, in MW_{th}	18.4	42.6	9.7	60
Electrical power produced by cycle, in MW_e	23.3	20.7	20.5	11.9
Parasitic losses, in MW_e	1.1	1.1	1.1	1.1
Net electrical power produced, in MW_e	22.2	19.6	19.4	10.8
Energy efficiency, in %	45.1	65.2	34.7	74.8
Electrical efficiency, in %	24.7	20.6	23.2	11.4
Exergy efficiency, in %	62.6	61.0	56.1	46.8

Simplified models for urban systems

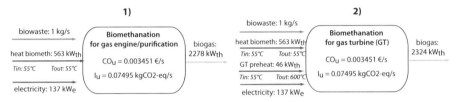

Figure D.1 Simplified model for the biomethanation of organic waste for the production of biogas 1) for a cogeneration engine or purification to SNG, 2) for a cogeneration turbine.

Figure D.2 Simplified model for the boilers using 1) woody biomass, 2) light fuel oil, 3) fossil natural gas, 4) biogenic SNG.

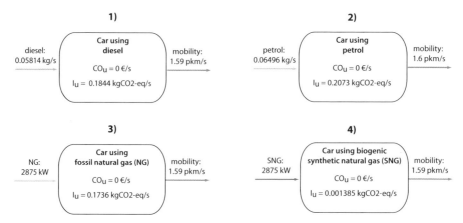

Figure D.3 Simplified model for the cars using 1) diesel, 2) petrol, 3) fossil natural gas, 4) biogenic SNG.

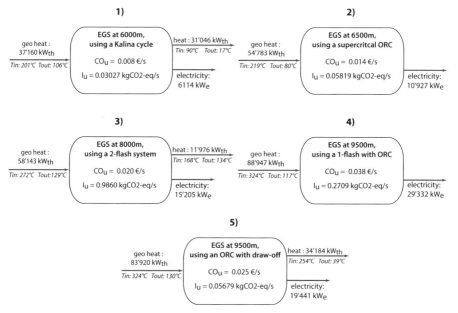

Figure D.4 Simplified model for the cycles used for the conversion of geothermal heat from EGS 1) Kalina cycle EGS, 2) supercritical ORC, 3) 2-flash, 4) 1-flash and bottoming ORC, 5) ORC with draw-off.

Figure D.5 Simplified model for the wood biomass dryers using 1) air, 2) steam.

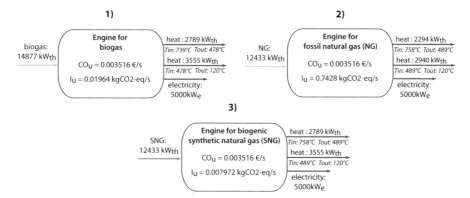

Figure D.6 Simplified model for the gas engines using 1) biogas, 2) fossil natural gas, 3) biogenic SNG.

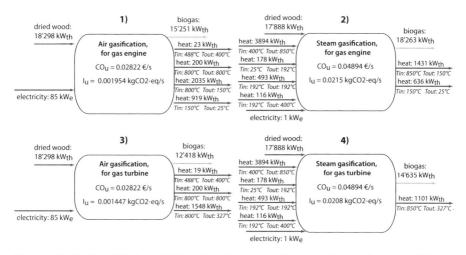

Figure D.7 Simplified model for the biomass gasifiers for the production of biogas 1) air gasification for engine, 2) steam gasification for engine, 3) air gasification for turbine, 4) steam gasification for turbine.

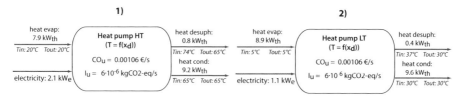

Figure D.8 Simplified model for the biomass gasifiers for the heat pumps 1) at high temperature (HT), 2) at low temperature (LT). The temperatures are only for the nominal operating conditions, and are given as a decision variable (x_d) in the MOO MINLP master problem.

Figure D.9 Simplified model for the MSWI plant. The electricity and district heating production are only for the nominal operating conditions, and their ratio is given as a decision variable (x_d) in the MOO MINLP master problem.

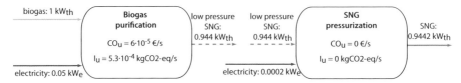

Figure D.10 Simplified model for the purification of biogas to synthetic natural gas (SNG) and its pressurization to meet the quality standards of the gas grid.

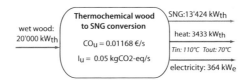

Figure D.11 Simplified model for the thermochemical conversion of woody biomass to synthetic natural gas (SNG).

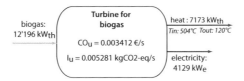

Figure D.12 Simplified model for the gas turbine using biogas.

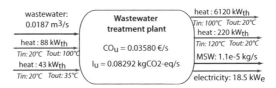

Figure D.13 Simplified model for the wastewater treatment plant.

Configurations for urban systems

Current situation

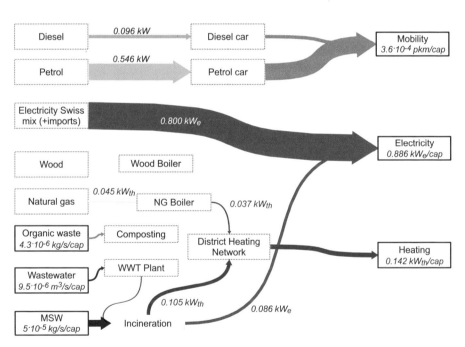

Figure E.1 Diagram representing the current system operation per capita for the city of La Chaux-dc-Fonds during period 1 (May, July & August).

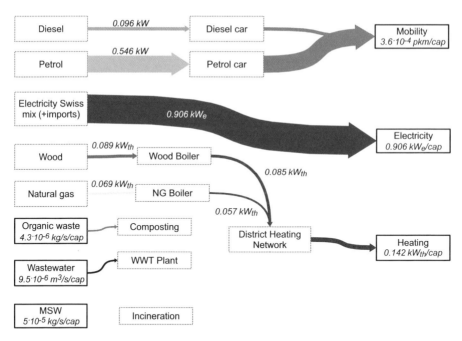

Figure E.2 Diagram representing the current system operation per capita for the city of La Chaux-de-Fonds during period 2 (June).

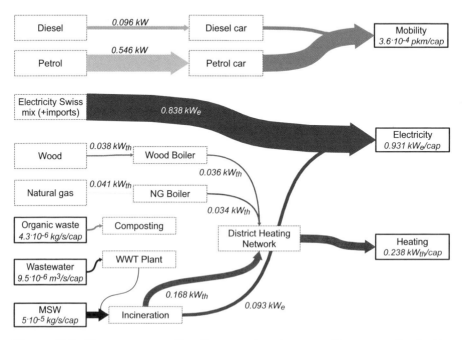

Figure E.3 Diagram representing the current system operation per capita for the city of La Chaux-de-Fonds during period 3 (April & September).

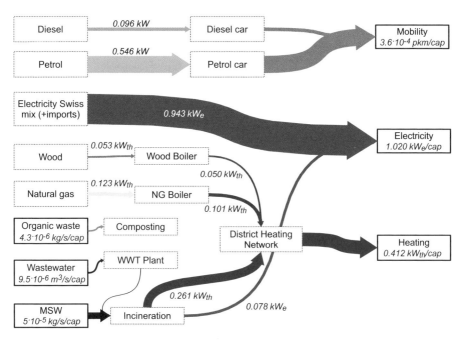

Figure E.4 Diagram representing the current system operation per capita for the city of La Chaux-de-Fonds during period 4 (March & October).

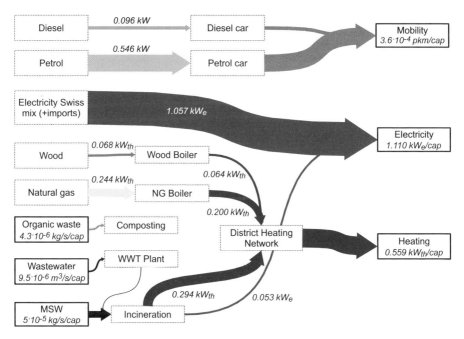

Figure E.5 Diagram representing the current system operation per capita for the city of La Chaux-de-Fonds during period 5 (November-February).

Configuration A1

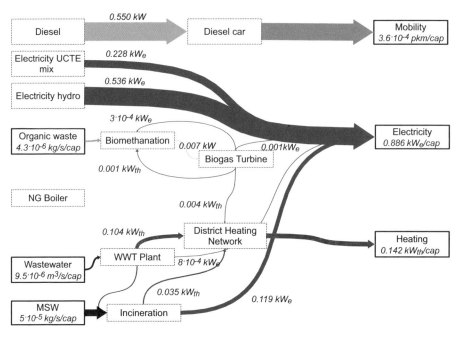

Figure E.6 Diagram representing the system operation per capita for the city of La Chaux-de-Fonds for configuration A1 during period 1 (May, July & August).

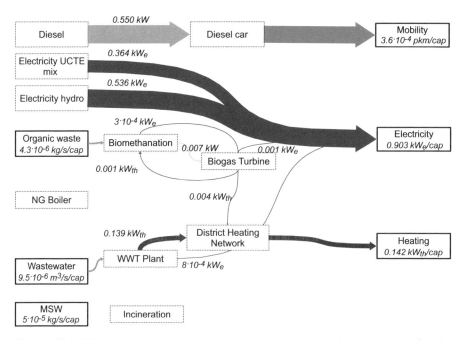

Figure E.7 Diagram representing the current system operation per capita for the city of La Chaux-de-Fonds for configuration A1 during period 2 (June).

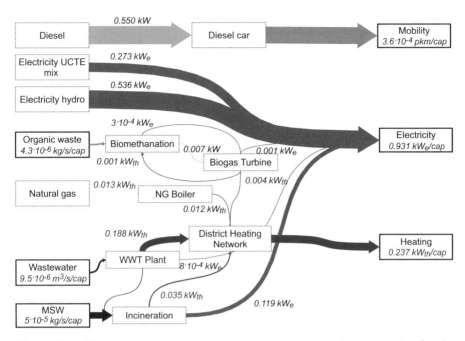

Figure E.8 Diagram representing the current system operation per capita for the city of La Chaux-de-Fonds for configuration A1 during period 3 (April & September).

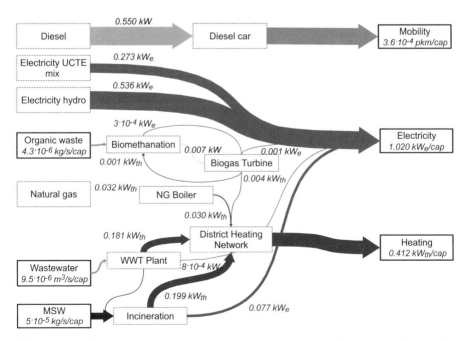

Figure E.9 Diagram representing the current system operation per capita for the city of La Chaux-de-Fonds for configuration A1 during period 4 (March & October).

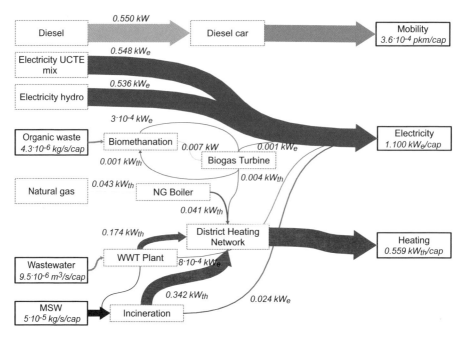

Figure E.10 Diagram representing the current system operation per capita for the city of La Chaux-de-Fonds for configuration A1 during period 5 (November-February).

Configuration A2

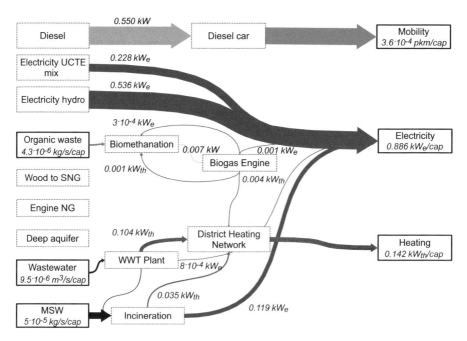

Figure E.11 Diagram representing the system operation per capita for the city of La Chaux-de-Fonds for configuration A2 during period 1 (May, July & August).

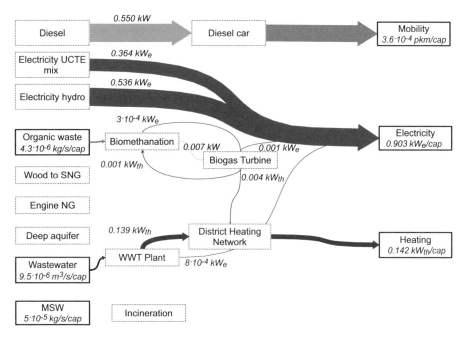

Figure E.12 Diagram representing the current system operation per capita for the city of La Chaux-de-Fonds for configuration A2 during period 2 (June).

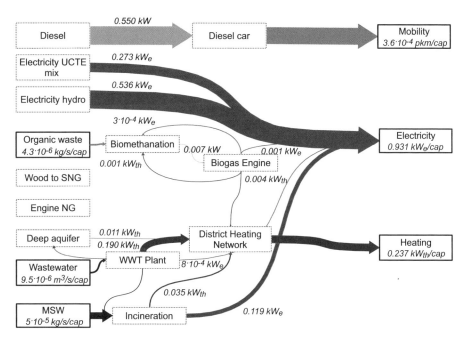

Figure E.13 Diagram representing the current system operation per capita for the city of La Chaux-de-Fonds for configuration A2 during period 3 (April & September).

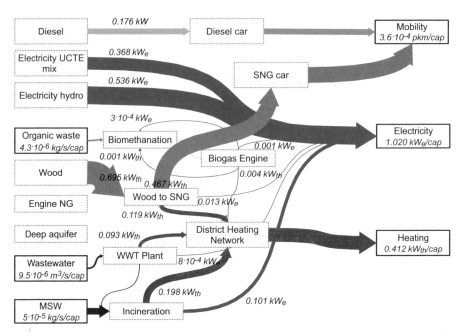

Figure E.14 Diagram representing the current system operation per capita for the city of La Chaux-de-Fonds for configuration A2 during period 4 (March & October).

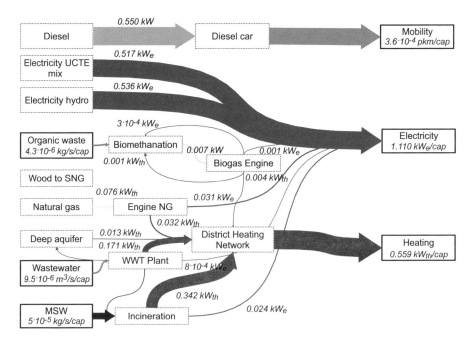

Figure E.15 Diagram representing the current system operation per capita for the city of La Chaux-de-Fonds for configuration A2 during period 5 (November-February).

Configuration B1

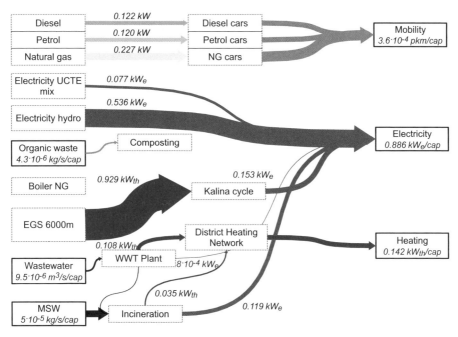

Figure E.16 Diagram representing the system operation per capita for the city of La Chaux-de-Fonds for configuration B1 during period 1 (May, July & August).

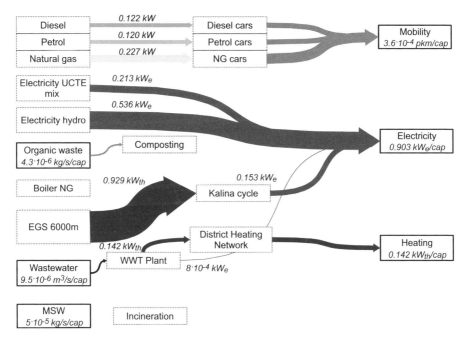

Figure E.17 Diagram representing the current system operation per capita for the city of La Chaux-de-Fonds for configuration B1 during period 2 (June).

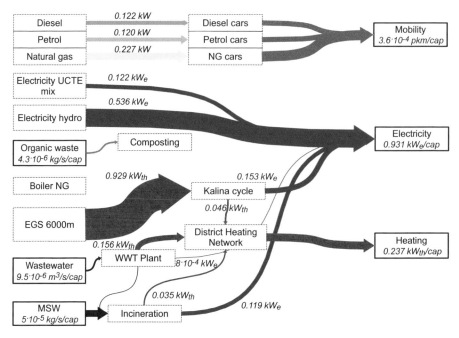

Figure E.18 Diagram representing the current system operation per capita for the city of La Chaux-de-Fonds for configuration B1 during period 3 (April & September).

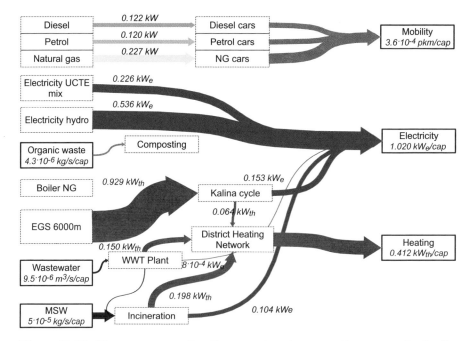

Figure E.19 Diagram representing the current system operation per capita for the city of La Chaux-de-Fonds for configuration B1 during period 4 (March & October).

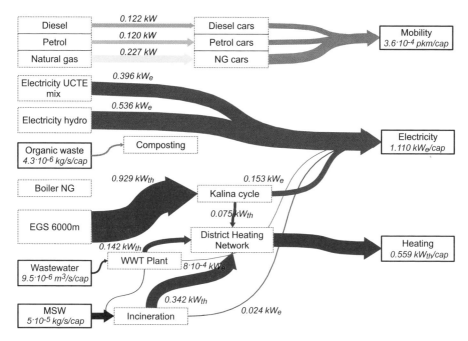

Figure E.20 Diagram representing the current system operation per capita for the city of La Chaux-de-Fonds for configuration B1 during period 5 (November-February).

Configuration B2

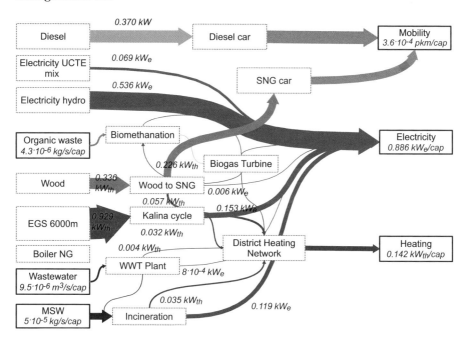

Figure E.21 Diagram representing the system operation per capita for the city of La Chaux-de-Fonds for configuration B2 during period 1 (May, July & August).

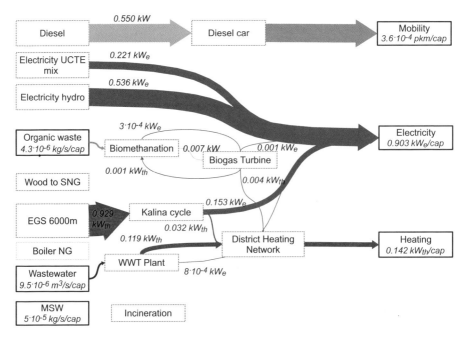

Figure E.22 Diagram representing the current system operation per capita for the city of La Chaux-de-Fonds for configuration B2 during period 2 (June).

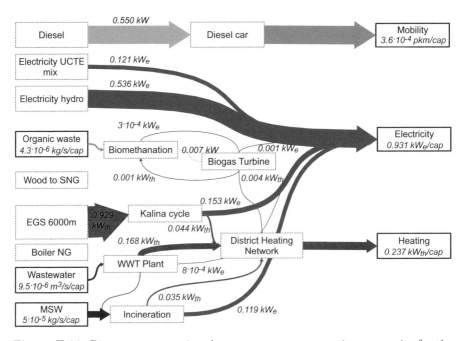

Figure E.23 Diagram representing the current system operation per capita for the city of La Chaux-de-Fonds for configuration B2 during period 3 (April & September).

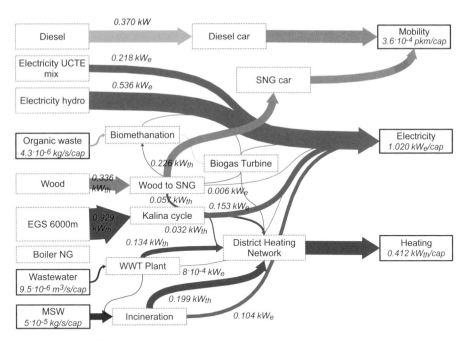

Figure E.24 Diagram representing the current system operation per capita for the city of La Chaux-de-Fonds for configuration B2 during period 4 (March & October).

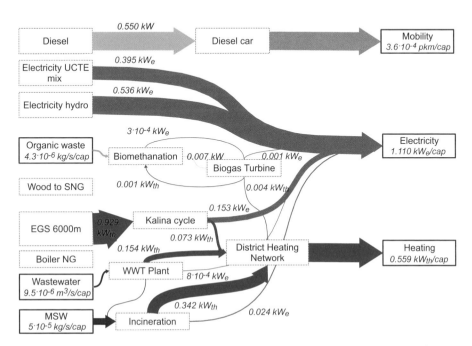

Figure E.25 Diagram representing the current system operation per capita for the city of La Chaux-de-Fonds for configuration B2 during period 5 (November-February).

Configuration C1

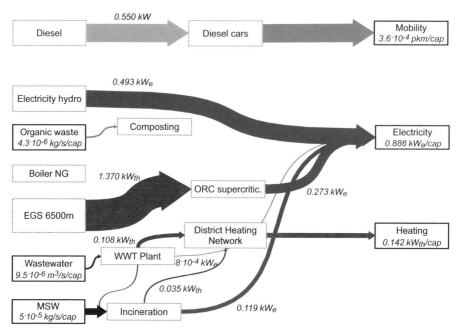

Figure E.26 Diagram representing the system operation per capita for the city of La Chaux-de-Fonds for configuration C1 during period 1 (May, July & August).

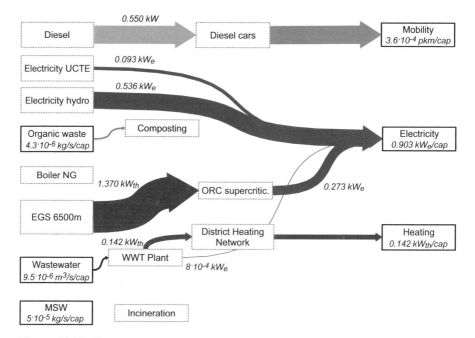

Figure E.27 Diagram representing the current system operation per capita for the city of La Chaux-de-Fonds for configuration C1 during period 2 (June).

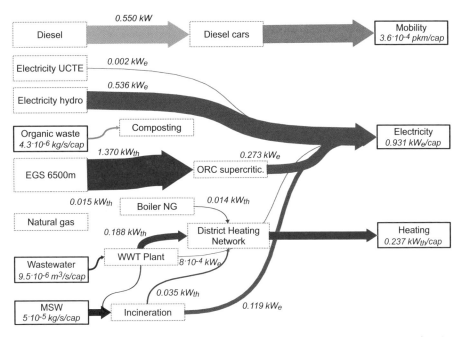

Figure E.28 Diagram representing the current system operation per capita for the city of La Chaux-de-Fonds for configuration C1 during period 3 (April & September).

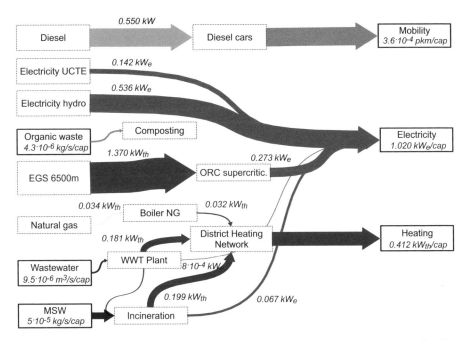

Figure E.29 Diagram representing the current system operation per capita for the city of La Chaux-de-Fonds for configuration C1 during period 4 (March & October).

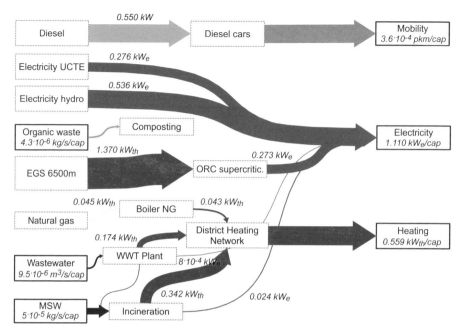

Figure E.30 Diagram representing the current system operation per capita for the city of La Chaux-de-Fonds for configuration C1 during period 5 (November-February).

Configuration C2

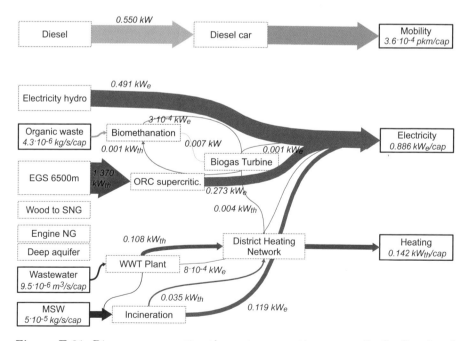

Figure E.31 Diagram representing the system operation per capita for the city of La Chaux-de-Fonds for configuration C2 during period 1 (May, July & August).

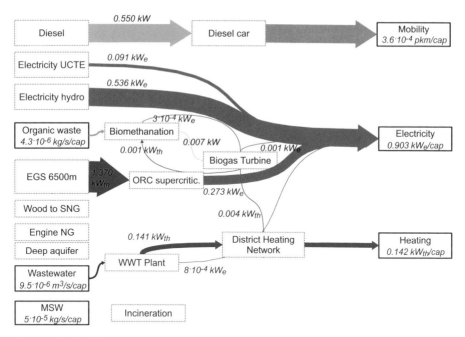

Figure E.32 Diagram representing the current system operation per capita for the city of La Chaux-de-Fonds for configuration C2 during period 2 (June).

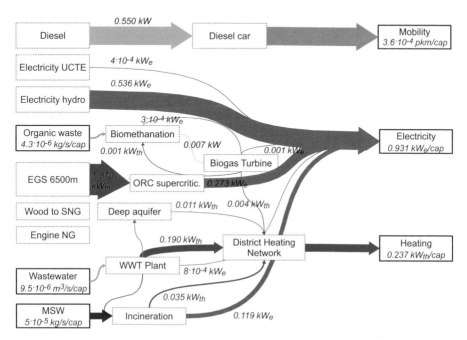

Figure E.33 Diagram representing the current system operation per capita for the city of La Chaux-de-Fonds for configuration C2 during period 3 (April & September).

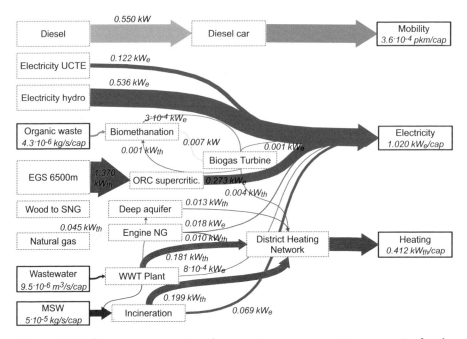

Figure E.34 Diagram representing the current system operation per capita for the city of La Chaux-de-Fonds for configuration C2 during period 4 (March & October).

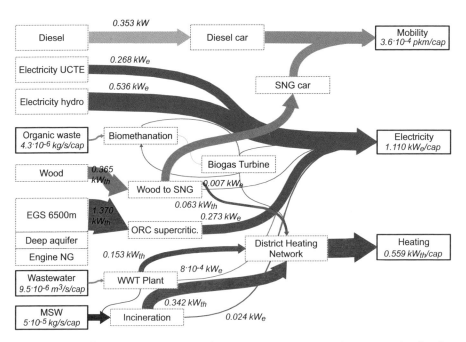

Figure E.35 Diagram representing the current system operation per capita for the city of La Chaux-de-Fonds for configuration C2 during period 5 (November-February).

Configuration D1

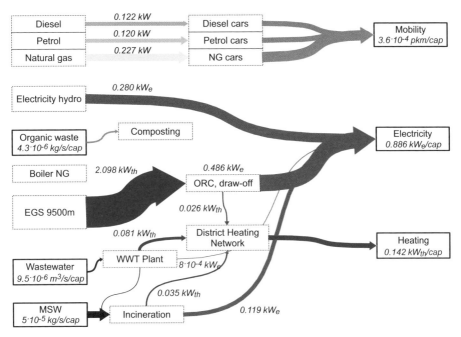

Figure E.36 Diagram representing the system operation per capita for the city of La Chaux-de-Fonds for configuration D1 during period 1 (May, July & August).

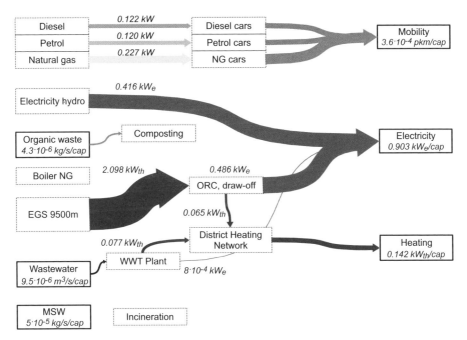

Figure E.37 Diagram representing the current system operation per capita for the city of La Chaux-de-Fonds for configuration D1 during period 2 (June).

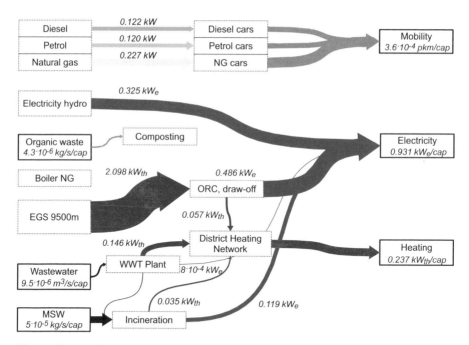

Figure E.38 Diagram representing the current system operation per capita for the city of La Chaux-de-Fonds for configuration D1 during period 3 (April & September).

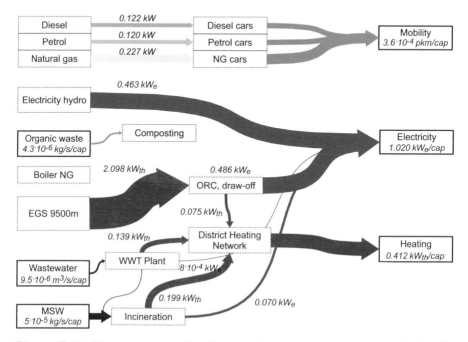

Figure E.39 Diagram representing the current system operation per capita for the city of La Chaux-de-Fonds for configuration D1 during period 4 (March & October).

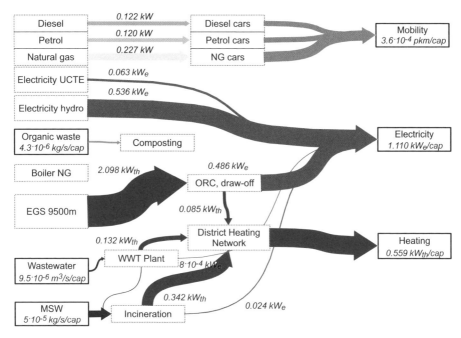

Figure E.40 Diagram representing the current system operation per capita for the city of La Chaux-de-Fonds for configuration D1 during period 5 (November-February).

Configuration D2

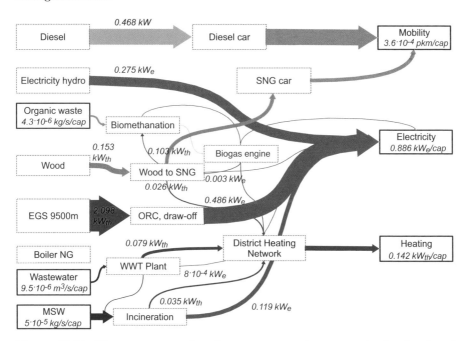

Figure E.41 Diagram representing the system operation per capita for the city of La Chaux-de-Fonds for configuration D2 during period 1 (May, July & August).

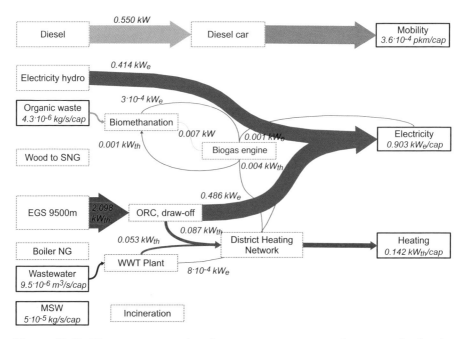

Figure E.42 Diagram representing the current system operation per capita for the city of La Chaux-de-Fonds for configuration D2 during period 2 (June).

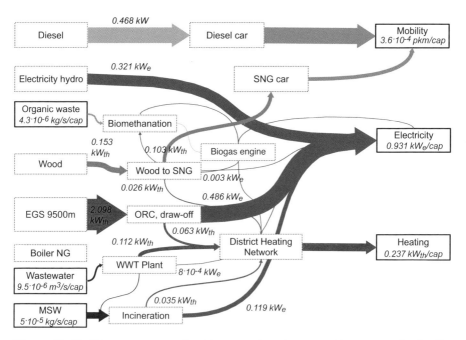

Figure E.43 Diagram representing the current system operation per capita for the city of La Chaux-de-Fonds for configuration D2 during period 3 (April & September).

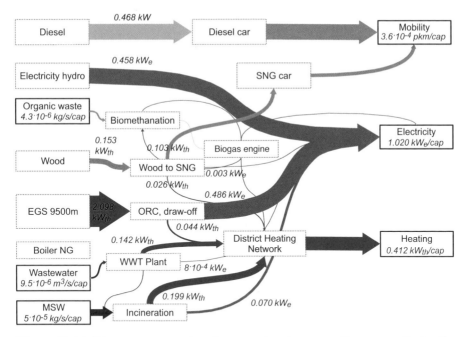

Figure E.44 Diagram representing the current system operation per capita for the city of La Chaux-de-Fonds for configuration D2 during period 4 (March & October).

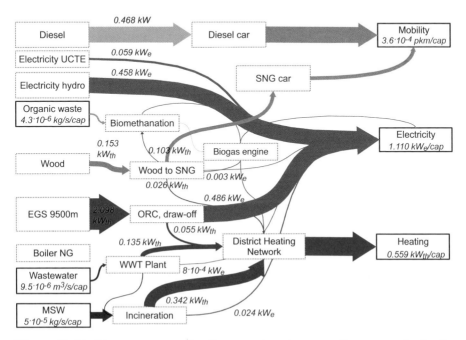

Figure E.45 Diagram representing the current system operation per capita for the city of La Chaux-de-Fonds for configuration D2 during period 5 (November-February).

Configuration E1

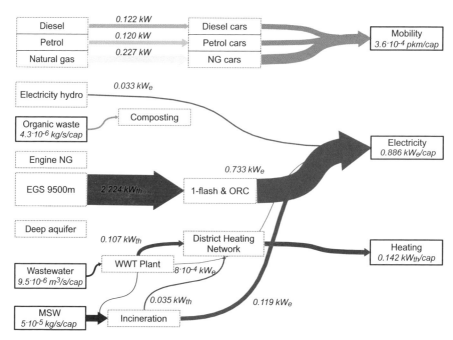

Figure E.46 Diagram representing the system operation per capita for the city of La Chaux-de-Fonds for configuration E1 during period 1 (May, July & August).

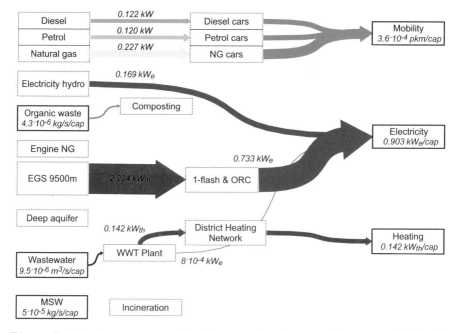

Figure E.47 Diagram representing the current system operation per capita for the city of La Chaux-de-Fonds for configuration E1 during period 2 (June).

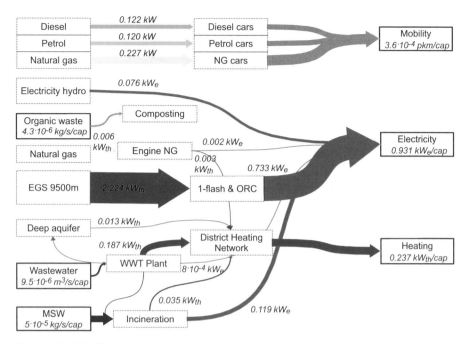

Figure E.48 Diagram representing the current system operation per capita for the city of La Chaux-de-Fonds for configuration E1 during period 3 (April & September).

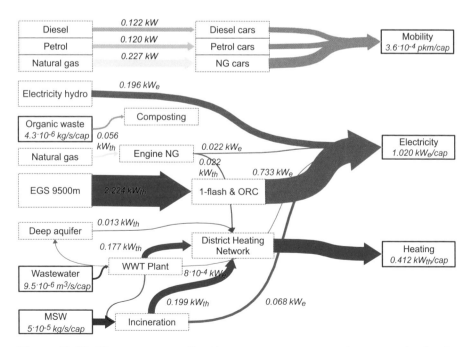

Figure E.49 Diagram representing the current system operation per capita for the city of La Chaux-de-Fonds for configuration E1 during period 4 (March & October).

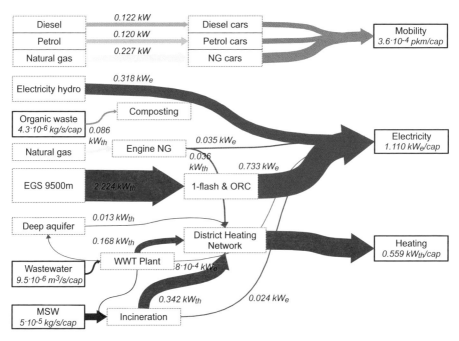

Figure E.50 Diagram representing the current system operation per capita for the city of La Chaux-de-Fonds for configuration E1 during period 5 (November-February).

Configuration E2

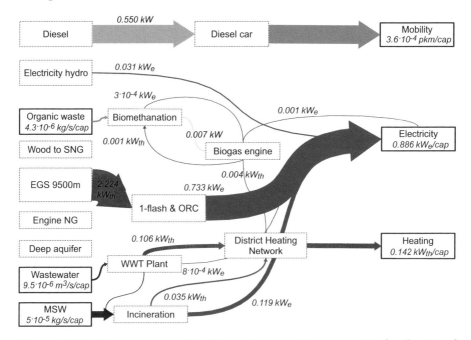

Figure E.51 Diagram representing the system operation per capita for the city of La Chaux-de-Fonds for configuration E2 during period 1 (May, July & August).

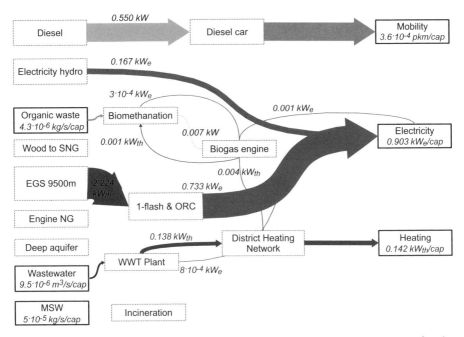

Figure E.52 Diagram representing the current system operation per capita for the city of La Chaux-de-Fonds for configuration E2 during period 2 (June).

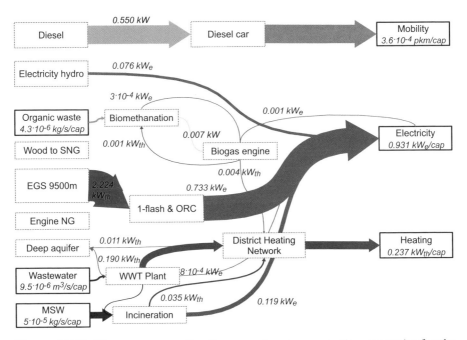

Figure E.53 Diagram representing the current system operation per capita for the city of La Chaux-de-Fonds for configuration E2 during period 3 (April & September).

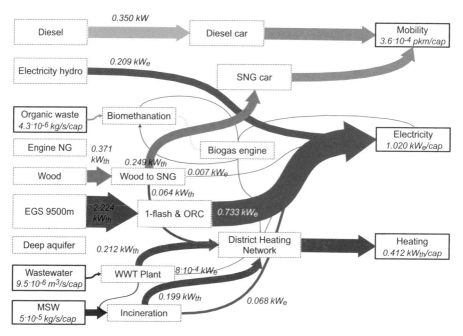

Figure E.54 Diagram representing the current system operation per capita for the city of La Chaux-de-Fonds for configuration E2 during period 4 (March & October).

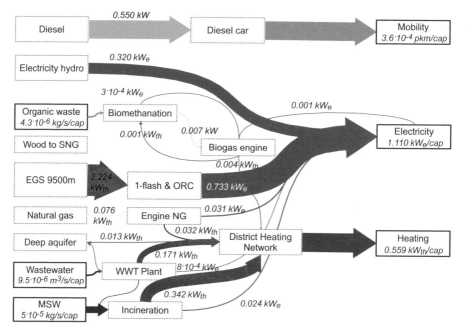

Figure E.55 Diagram representing the current system operation per capita for the city of La Chaux-de-Fonds for configuration E2 during period 5 (November-February).

Bibliography

ALEXANDER B., BARTON G., PETRIE J., ROMAGNOLI J., Process synthesis and optimisation tools for environmental design: methodology and structure, *Computers & Chemical Engineering*, vol. 24, 1195–1200, 2000.

ALLENBY B., RICHARDS D., *The Greening of Industrial Ecosystems*, National Academy Press, Washington D.C, 1994.

ALTHAUS H., BLASER S., CLASSEN M., JUNGBLUTH N., *Life Cycle Inventories of Metals, ecoinvent report No 10*, Tech. rep., ecoinvent center, CH-8600 Dübendorf, Switzerland, 2007a.

ALTHAUS H., CHUDACOFF M., HISCHIER R., JUNGBLUTH N., OSSES M., PRIMAS A. *Life Cycle Inventories of Chemicals, ecoinvent report No 8*, Tech. rep., ecoinvent center, CH-8600 Dübendorf, Switzerland, 2007b.

ALTHAUS H., DINKEL F., WERNER F., *Life Cycle Inventories of Renewable Materials, ecoinvent report No 21*, Tech. rep., ecoinvent center, CH-8600 Dübendorf, Switzerland, 2007c.

AYRES R., Life cycle analysis: A critique, *Resources, Conservation and Recycling*, vol. 14, 199–223, 1993.

AZAPAGIC A., CLIFT R., Life Cycle Assessment and Linear Programming - Environmental optimisation of a product system, *Computers & Chemical Engineering*, vol. 19, 229–234, 1995.

AZAPAGIC A., CLIFT R., The application of life cycle assessment to process optimization, *Computers & Chemical Engineering*, vol. 23, 1509–1526, 1999.

BALDACCI A., BRUNAZZI E., GALLETTI C., PAGLIANTI A., Modelling and experimental validation of H2S emissions in geothermal power plants, *Geothermics*, vol. 31, 501–517, 2002.

BECKER H., MARÉCHAL F., Energy integration of industrial sites with heat exchange restrictions, *Computers & Chemical Engineering*, vol. 37, 104–118, 2012a.

BECKER H., MARÉCHAL F., Targeting industrial heat pump integration in multiperiod problems, *Proceedings of the 11th International Symposium on Process Systems Engineering*, 2012b.

BELSIM, www.belsim.com, official website of the Belsim company, Awans, Belgium, last date of access: 18.10.2011, 2011.

BERKEL R. V., FUJITA T., HASHIMOTO S., GENG Y., Industrial and urban symbiosis in Japan: Analysis of the Eco-Town program 1997-2006, *Journal of Environmental Management*, vol. 90, 1544–1556, 2009.

BERNIER E., MARÉCHAL F., SAMSON R., Multi-objective design optimisation of a natural gas-combined cycle with carbon dioxide capture in a life cycle perspective, *Energy*, vol. 35, 1121–1128, 2010.

BOLLIGER R., *Méthodologie de la synthèse des systèmes énergétiques industriels*, Ph.D. thesis, Ecole Polytechnique Fédérale de Lausanne, 2010.

BOREL L., FAVRAT D., *Thermodynamique et Energétique: Volume 1, de l'énergie à l'exergie*, Presses polytechniques et universitaires romandes, EPFL - Centre Midi, CH-1015 Switzerland, 2005.

BRAND G., BRAUNSCHWEIG A., SCHEIDEGGER A., SCHWANK O., *Bewertung in Oekobilanzen mit der Methode des oekologischen Knappheit Oekofaktoren 1997*, Tech. rep., Swiss Federal Office of Environment, Bern, Switzerland, 1998.

BROWN M., ULGIATI S., Emergy evaluations and environmental loading of electricity production systems, *Journal of Cleaner Production*, vol. 10, 321–334, 2002.

CHERTOW M., Industrial Symbiosis: Literature and taxonomy, *Annual Review of Energy and the Environment*, vol. 25, 313–337, 2000.

CHERTOW M., LOMBARDI D., Quantifying Economic and Environmental Benefits of Co-Located Firms, *Environmental Science and Technology*, vol. 39, 6535–6541, 2005.

CIMREN E., FIKSEL J., POSNER M., SIKDAR K., Material Flow Optimization in By-product Synergy Network, *Journal of Industrial Ecology*, vol. 15, 315–332, 2012.

CORNELISSEN R., HIRS G., The value of the exergetic life cycle assessment besides the LCA, *Energy Conversion and Management*, vol. 43, 1417–1424, 2002.

CRIDOR, *Information booklet "Nos déchets - Notre Energie"* published by the Centre de Valorisation des déchets Arc jurassien (Cridor), available at: www.cridor.ch, 2011.

CUDILLEIRO M., *Integration of LCA in the thermo-economic models of BtL processes*, Master thesis effectuated at Ecole Polytechnique Fédérale de Lausanne, 2010.

CUENOT N., FAUCHER J.-P., FRITSCH D., GENTER A., SZABLINSKI D., The European EGS project at Soultz-sous-Forêts: from extensive exploration to power production, *IEEE Power and Energy Society 2008 General Meeting: Conversion and Delivery of Electrical Energy in the 21st Century*, PES, art. no. 4596680, 2008.

CUVILLIEZ A.-L., *Life Cycle Analysis of a thermo-economic model of a lignocellulosic bioethanol process*, Master thesis effectuated at Ecole Polytechnique Fédérale de Lausanne and at Cornell University, 2009.

DEDIEGO L., LONDONOT C., WANG X., GIBBS B., Influence of operating parameters on NOx and N2O axial profiles in a circulating fluidized bed combustor, *Fuel*, vol. 75, 971–978, 1996.

DESCOINS N., DELERIS S., LESTIENNE R., TROUVÉ E., MARÉCHAL F., Energy efficiency in waste water treatment plants: Optimization of activated sludge process coupled with anaerobic digestion, *Energy*, vol. 41, 153–164, 2012.

DESIDERI U., BIDINI G., Study of possible optimisation criteria for geothermal power plants, *Energy Conversion and Management*, vol. 38, 1681–1691, 1997.

DiPippo R., Second Law assessment of binary plants generating power from low-temperature geothermal fluids, *Geothermics*, vol. 33, 565–586, 2004.

DiPippo R., *Geothermal Power Plants: Principles, Applications, Case Studies and Environmental Impacts*, Butterworth-Heinemann c/o Elsevier, Oxford, UK, 2008.

Diwekar U., Shastri Y., Green process design, green energy, and sustainability: A systems analysis perspective, *Computers and Chemical Engineering*, vol. 34, 1348–1355, 2010.

Doka G., *Life Cycle Inventories of Waste Treatment Services, ecoinvent report No 13*, Tech. rep., ecoinvent center, CH-8600 Dübendorf, Switzerland, 2007.

Dones R., Bauer C., *Life Cycle Inventories of Transport Services, ecoinvent report No 14*, Tech. rep., ecoinvent center, CH-8600 Dübendorf, Switzerland, 2007.

Dones R., Bauer C., Bolliger R., Burger B., Heck T., Ršder A., Emmenegger M. F., Frischknecht R., Jungbluth N., Tuchschmid M., *Sachbilanzen von Energiesystemen, ecoinvent report No 6*, Tech. rep., ecoinvent center, CH-8600 Dübendorf, Switzerland, 2007.

Ehrenfeld J., Industrial ecology: A framework for product and process design, *Journal of Cleaner Production*, vol. 5, 87–95, 1997.

Ehrenfeld J., Gertler N., Industrial Ecology in Practice. The Evolution of Interdependence at Kalundborg, *Journal of Industrial Ecology*, vol. 1, 67–79, 1997.

Erkman S., Industrial ecology: an historical view, *Journal of Cleaner Production*, vol. 5, 1–10, 1997.

Evans A., Strezov V., Evans T., Assessment of sustainability indicators for renewable energy technologies, *Renewable and Sustainable Energy Reviews*, vol. 13, 1082–1088, 2009.

Fazlollahi S., Maréchal F., Multi-objective, multi-period optimization of biomass conversion technologies using evolutionary algorithms and mixed integer linear programming (MILP), *Applied Thermal Engineering*, vol. 50, 1504–1513, 2013.

Felder R., *Ecological Impact of the Use of Methane from Wood Gasification, Project Ecogas, Internal Report*, Tech. rep., Paul Scherrer Institute (PSI), CH-5232 Villigen, Switzerland, 2004.

Felder R., Dones R., Evaluation of ecological impacts of synthetic natural gas from wood used in current heating and car systems, *Biomass and bioenergy*, vol. 31, 403–415, 2007.

Finnveden G., On the limitations of life cycle assessment and environmental systems analysis tools in general, *International Journal of Life Cycle Assessment*, vol. 5, 229–238, 2000.

Finnveden G., Hauschild M., Ekvall T., Guinée J., Heijungs R., Hellweg S., Koehler A., Pennington D., Suh S., Recent developments in Life Cycle Assessment, *Journal of Environmental Management*, vol. 91, 1–21, 2009.

Floudas C. A., Ciric A., Grossmann I., Automatic Synthesis of Optimum Heat Exchanger Network Configurations, *AIChE Journal*, vol. 32, 276–290, 1986.

FRANCO A., VILLANI M., Optimum design of binary cycle power plants for water-dominated, medium-temperature geothermal fields, *Geothermics*, vol. 38, 379–391, 2009.

FRANGOPOULOS C., Thermo-economic functional analysis and optimization, *Energy*, vol. 12, 563–571, 1987.

FRICK S., KALTSCHMITT M., SCHRÖDER G., Life cycle assessment of geothermal binary power plants using enhanced low-temperature reservoirs, *Energy*, vol. 35, 2281–2294, 2010.

FRISCHKNECHT R., JUNGBLUTH N., *Overview and Methodology, ecoinvent report No 1*, Tech. rep., ecoinvent center, CH-8600 Dübendorf, Switzerland, 2007.

FRISCHKNECHT R., JUNGBLUTH N., ALTHAUS H.-J., DOKA G., DONES R., HECK T., HELLWEG S., HISCHIER R., NEMECEK T., REBITZER G., SPIELMANN M., The ecoinvent database: Overview and Methodological Framework, *International Journal of Life Cycle Assessment*, vol. 10, 3–9, 2005.

FRONDINI F., CALIRO S., CARDELLINI C., CHIODINI G., MORGANTINI N., Carbon dioxide degassing and thermal energy release in the Monte Amiata volcanic-geothermal area (Italy), *Applied Geochemistry*, vol. 24, 860–875, 2009.

GASSNER M., *Process Design Methodology for Thermochemical Production of Fuels from Biomass. Application to the Production of Synthetic Natural Gas from Lignocellulosic Resources*, Ph.D. thesis, Ecole Polytechnique Fédérale de Lausanne, 2010.

GASSNER M., GERBER L., SALGUEIRO L., MARÉCHAL F., *Perspectives de l'utilisation du biogaz distribué dans le réseau de gaz naturel* - Rapport de l'étude effectuée sur mandat du Service de l'Environnement et de l'Energie (SEVEN) du Canton de Vaud, Tech. rep., Ecole Polytechnique Fédérale de Lausanne (EPFL), CH-1015 Lausanne, Switzerland, 2011.

GASSNER M., MARÉCHAL F., Methodology for the optimal thermo-economic, multi-objective design of thermochemical fuel production from biomass, *Computers & Chemical Engineering*, vol. 33, 769–781, 2009a.

GASSNER M., MARÉCHAL F., Thermo-economic process model for thermochemical production of Synthetic Natural Gas (SNG) from lignocellulosic biomass, *Biomass and bioenergy*, vol. 33, 1587–1604, 2009b.

GENTER A., CUENOT N., GOERKE X., MELCHERT B., SANJUAN B., SCHEIBER J., Status of the Soultz Geothermal Project During Exploitation Between 2010 and 2012, *Proceedings of the 37th Workshop on Geothermal Reservoir Engineering*, pp. 704–715, 2012.

GERBER L., FAZLOLLAHI S., MARÉCHAL F., Environomic optimal design and synthesis of energy conversion systems in urban areas, *Computer Aided Chemical Engineering*, vol. 30, 41–45, 2012.

GERBER L., FAZLOLLAHI S., MARÉCHAL F., A systematic methodology for the environomic design and synthesis of energy systems combining process integration, Life Cycle Assessment and industrial ecology, *Computers & Chemical Engineering*, vol. 59, 2–16, 2013.

GERBER L., GASSNER M., MARÉCHAL F., Systematic Integration of LCA in process systems design: Application to combined fuel and electricity production from ligno-cellulosic biomass, *Computers & Chemical Engineering*, vol. 35, 1265–1280, 2011a.

GERBER L., MARÉCHAL F., *Systèmes hybrides pour les installations de géothermie profondes: cas d'étude de La Chaux-de-Fonds* - Rapport de l'étude effectuée sur mandat du Laboratoire de Géothermie (CREGE) de l'Université de Neuchâtel, Tech. rep., Ecole Polytechnique Fédérale de Lausanne (EPFL), CH-1015 Lausanne, Switzerland, 2011.

GERBER L., MARÉCHAL F., Defining optimal configurations of geothermal systems using process design and process integration techniques, *Applied Thermal Engineering*, vol. 43, 29–41, 2012a.

GERBER L., MARÉCHAL F., Design of geothermal energy conversion systems with a life cycle assessment perspective, *Proceedings of the 37th Stanford Geothermal Workshop*, January 30 - February 1, 2012b.

GERBER L., MARÉCHAL F., Environomic optimal configurations of geothermal energy conversion systems: application to the future construction of Enhanced Geothermal Systems in Switzerland, *Energy*, vol. 45, 908–923, 2012c.

GERBER L., MAYER J., MARÉCHAL F., A systematic methodology for the synthesis of unit process chains using Life Cycle Assessment and Industrial Ecology Principles, *Computer Aided Chemical Engineering*, vol. 29, 1215–1219, 2011b.

GIRARDIN L., MARÉCHAL F., DUBUIS M., CALAME-DARBELLAY N., FAVRAT D., EnerGis: A geographical information based system for the evaluation of integrated energy conversion systems in urban areas, *Energy*, vol. 35, 830–840, 2010.

GOEDKOOP M., SPRIENSMA R., *The Eco-Indicator 99: A damage oriented method for life cycle impact assessment*, Tech. rep., PRé Consultants, Amersfoort, The Netherlands, 2000.

GROSSMANN I., GUILLÉN-GOSÁLBEZ G., Scope for the application of mathematical programming techniques in the synthesis and planning of sustainable processes, *Computers & Chemical Engineering*, vol. 34, 1365–1376, 2010.

GUILLÉN-GOSÁLBEZ G., CABALLERO J., JIMÉNEZ L., Application of Life Cycle Assessment to the Structural Optimization of Process Flowsheets, *Industrial and Engineering Chemistry Research*, vol. 47, 777–789, 2008.

GUILLÉN-GOSÁLBEZ G., GROSSMANN I., A global optimization strategy for the environmentally conscious design of chemical supply chains under uncertainty in the damage assessment model, *Computers & Chemical Engineering*, vol. 34, 42–58, 2010.

GUINÉE J., HEIJUNGS R., DE HAES H. U., HUPPES G., Quantitative Life Cycle Asssessment of products - 1: Goal definition and inventory, *Journal of Cleaner Production*, vol. 1, 3–91, 1993a.

GUINÉE J., HEIJUNGS R., DE HAES H. U., HUPPES G., Quantitative Life Cycle Asssessment of products - 2: Classification, valuation and improvement analysis, *Journal of Cleaner Production*, vol. 1, 81–13, 1993b.

GUINÉE J., HEIJUNGS R., DE HAES H. U., HUPPES G., Quantitative Life Cycle Asssessment of products - 2: Classification, valuation and improvement analysis, *Journal of Cleaner Production*, vol. 1, 131–137, 1993c.

GUINÉE J., HEIJUNGS R., HUPPES G., ZAMAGNI A., MASONI P., BUONAMICI R., EKVALL T., RYDBERG T., Life Cycle Assessment: Past, Present and Future, *Environmental Science and Technology*, vol. 45, 90–96, 2011.

GUO T., WANG H., ZHANG S., Selection of working fluids for a novel low-temperature geothermally-powered ORC based cogeneration system, *Energy Conversion and Management*, vol. 52, 2384–2391, 2011.

HARING M., Deep heat mining: Development of a cogeneration power plant from an enhanced geothermal system in Basel, Switzerland, *Geothermal Resources Council Transactions*, vol. 28, 219–222, 2004.

HETTIARACHCHI H., GOLUBOVIC M., WOREK W., IKEGAMI Y., Optimum design criteria for an Organic Rankine cycle using low-temperature geothermal heat sources, *Energy*, vol. 32, 1698–1706, 2007.

HISCHIER R., *Life Cycle Inventories of Packaging and Graphical Paper, ecoinvent report No 11*, Tech. rep., ecoinvent center, CH-8600 Dübendorf, Switzerland, 2007.

HISCHIER R., CLASSEN M., LEHMANN M., SCHARNHORST W., *Life Cycle Inventories of Electric and Electronic Equipment - Production, Use & Disposal, ecoinvent report No 18*, Tech. rep., ecoinvent center, CH-8600 Dübendorf, Switzerland, 2007.

HOFBAUER H., RAUCH R., *Hydrogen-rich gas from biomass steam gasification*, publishable final report, Tech. rep., Institute of Chemical Engineering Fuel and Environmental Technology, A-1060 Vienna, Austria, 2001.

HUGO A., PISTIKOPOULOS E., Environmentally conscious long-range planning and design of supply chain networks, *Journal of Cleaner Production*, vol. 13, 1471–1491, 2005.

HÄSSIG W., PRIMAS A., *Life Cycle Inventories of Comfort Ventilation in Dwellings, ecoinvent report No 25*, Tech. rep., ecoinvent center, CH-8600 Dübendorf, Switzerland, 2007.

IEA, *Technology Roadmap - Geothermal Heat and Power*, Tech. rep., International Energy Agency, Paris, France, 2011.

INTERGOVERNMENTAL PANEL ON CLIMATE CHANGE, *IPCC(2007) Climate Change 2007: The Scientific Basis. In: Fourth Assessment Report of the Intergovernmental Panel on Climate Change (2007)*, Tech. rep., IPCC, 2007.

ISO, *Environmental management, Life Cycle Assessment, Principles and Framework. International Standard, ISO 14 040*, 2006a.

ISO, *Environmental management, Life Cycle Assessment, Requirements and Guidelines. International Standard, ISO 14 044*, 2006b.

JACOBSEN N., Industrial Symbiosis in Kalundborg, Denmark - A Quantitative Assessment of Economic and Environmental Aspects, *Journal of Industrial Ecology*, vol. 10, 239–255, 2006.

JUNGBLUTH N., CHUDACOFF M., DAURIAT A., DINKEL F., DOKA G., EMMENEGGER M. F., GNANSOUNOU E., KLJUN N., SPIELMANN M., STETTLER C., SUTTER J., *Life Cycle Inventories of Bioenergy, ecoinvent report No 17*, Tech. rep., ecoinvent center, CH-8600 Dübendorf, Switzerland, 2007.

KALTSCHMITT M., REINHARDT G., STELZER T., Life Cycle Analysis of biofuels under different environmental aspects, *Biomass and Bioenergy*, vol. 12, 121–134, 1997.

KANOGLU M., DINCER I., Performance assessment of cogeneration plants, *Energy Conversion and Management*, vol. 50, 76–81, 2009.

KELLENBERGER D., ALTHAUS H., JUNGBLUTH N., KÜNNIGER T., *Life Cycle Inventories of Building Products, ecoinvent report No 7*, Tech. rep., ecoinvent center, CH-8600 Dübendorf, Switzerland, 2007.

KEMP I., *Pinch Analysis and Process Integration: A User Guide on Process Integration for the Efficient Use of Energy*, Elsevier Butterworth-Heinemann, Amsterdam, Netherlands, 2007.

KEOLEIAN G., The application of life cycle assessment to design, *Journal of Cleaner Production*, vol. 1, 143–149, 1993.

KIRSCHNER M. J., Oxygen, *Ullmann's Encyclopedia of Industrial Chemistry*, 7th Edition, Wiley-VCH, 2009.

KNIEL G., DELMARCO K., PETRIE J., Life Cycle Assessment to Process Design: Environmental and Economic Analysis and Optimization of a Nitric Acid Plant, *Environmental Progress*, vol. 15, 221–228, 1996.

LAZZARETTO A., TOFFOLO A., Energy, economy and environment as objectives in multi-criterion optimization of thermal systems design, *Energy*, vol. 29, 1139–1157, 2004.

LAZZARETTO A., TOFFOLO A., MANENTE G., ROSSI N., PACI M., Cost Evaluation of organic Rankine cycles for Low temperature geothermal sources, *Proceedings of the 24th International Conference on Efficiency, Cost, Optimization, Simulation and Environmental Impact of Energy Systems*, 3854–3868, 2011.

LI H., BURER M., SONG Z.-P., FAVRAT D., MARÉCHAL F., Green heating system: characteristics and illustration with multi-criteria optimization of an integrated energy system, *Energy*, vol. 29, 225–244, 2004.

LI H., MARÉCHAL F., BURER M., FAVRAT D., Multi-objective optimization of an advanced combined cycle power plant including co2 separation systems, *Energy*, vol. 31, 3117–3134, 2006.

LINNHOFF B., TOWNSEND D., BOLAND D., HEWITT G., THOMAS B., GUY A., MARSLAND R., *User Guide on Process Integration for the Efficient Use of Energy*, The Institution of Chemical Engineers, Rugby, Warks, England, 1982.

LUTERBACHER J. S., FRÖLING M., VOGEL F., MARÉCHAL F., TESTER J. W., Hydrothermal gasification of waste biomass: Process design and life cycle assessment, *Environmental Science and Technology*, vol. 43, 1578–1583, 2009.

MAJER E., BARIA R., STARK M., OATES S., BOMMER J., SMITH B., ASANUMA H., Induced seismicity associated with Enhanced Geothermal Systems, *Geothermics*, vol. 36, 185–222, 2007.

MANN M., COLLINGS M., BOTROS P., Nitrous oxide emissions in fluidized bed combustion: fundamental chemistry and combustion testing, *Progress in Energy Combustion*, vol. 18, 447–461, 1992.

MARÉCHAL F., KALITVENTZEFF B., Targeting the minimum cost of energy requirements: A new graphical technique for evaluating the integration of utility systems, *Computers & Chemical Engineering*, vol. 20, S225–S230, 1996.

MARÉCHAL F., KALITVENTZEFF B., Process integration: Selection of the optimal utility system, *Computers & Chemical Engineering*, vol. 22, S149–S156, 1998.

MARTINEZ P., ELICECHE A., Minimization of life cycle CO2 emissions in steam and power plants, *Clean Technologies and Environmental Policy*, vol. 11, 49–57, 2009.

MATTILA T., LEHTORANTA S., SOKKA L., MELANEN M., NISSINEN A., Methodological Aspects of Applying Life Cycle Assessment to Industrial Symbioses, *Journal of Industrial Ecology*, vol. 16, 51–60, 2012.

MATTILA T., PAKARINEN S., SOKKA L., Quantifying the Total Environmental Impacts of an Industrial Symbiosis - a Comparison of Process-, Hybrid- and Input-Output Life Cycle Assessment, *Environmental Science and Technology*, vol. 44, 4309–4314, 2010.

MÉGEL O., *Analyse énergétique d'un quartier et étude d'un réseau de chauffage à distance, La Chaux-de-Fonds*, Master thesis effectuated at Ecole Polytechnique Fédérale de Lausanne, 2011.

MINDER R., KÖDEL J., SCHÄDLE K.-H., RAMSEL K., GIRARDIN L., MARÉCHAL F., *Energy conversion processes for the use of geothermal heat*, Tech. rep., Swiss Federal Office of Energy, Bern, Switzerland, 2007.

MLCAK H., Kalina Cycle Concepts for Low Temperature Geothermal, *Geothermal Resources Council Transactions*, 707–713, 2002.

MOLYNEAUX A., LEYLAND G., FAVRAT D., Environomic multi-objective optimisation of a district heating network considering centralized and decentralized heat pumps, *Energy*, vol. 35, 751–758, 2010.

MULLER E., KOBEL B., SCHMID F., *Energie dans les stations d'épuration*, Tech. rep., Suisse Energie (Swiss Federal Office of Energy), Bern, Switzerland, 2008.

OFEN, *Schweizerische Holzenergiestatistik*, Tech. rep., Swiss Federal Office of Energy, Bern, Switzerland, 2010a.

OFEN, *Statistique globale suisse de l'énergie 2010*, Tech. rep., Swiss Federal Office of Energy, Bern, Switzerland, 2010b.

OFEN, *Statistique suisse de l'électricité 2010*, Tech. rep., Swiss Federal Office of Energy, Bern, Switzerland, 2010c.

OFEV, *L'environnement suisse - Statistique de poche 2011*, Tech. rep., Swiss Federal Office of Environment, Bern, Switzerland, 2011.

ORMAT, Ormat Systems Ltd, e-mail communication from the 12/06/2010, 2010.

OSTAT, *Mobilité et transports 2010*, Tech. rep., Swiss Federal Office of Statistics, Bern, Switzerland, 2010.

OZGENER L., HEPBASLI A., DINCER I., ROSEN M., Exergoeconomic analysis of geothermal district heating systems: A case study, *Applied Thermal Engineering*, vol. 27, 1303–1310, 2007.

PAPANDREOU V., SHANG Z., A multi-criteria optimisation approach for the design of sustainable utility systems, *Computers & Chemical Engineering*, vol. 32, 1589–1602, 2008.

PAPOULIAS S., GROSSMANN I., A structural optimization approach in process synthesis - i. utility systems, *Computers & Chemical Engineering*, vol. 7, 695–706, 1983a.

PAPOULIAS S., GROSSMANN I., A structural optimization approach in process synthesis - ii. heat recovery networks, *Computers & Chemical Engineering*, vol. 7, 707–721, 1983b.

PAPOULIAS S., GROSSMANN I., A structural optimization approach in process synthesis - iii. total processing, *Computers & Chemical Engineering*, vol. 7, 723–734, 1983c.

PEHNT M., Dynamic life cycle assessment (LCA) of renewable energy technologies, *Renewable Energy*, vol. 31, 55–71, 2006.

PORTIER S., VUATAZ F.-D., NAMI P., SANJUAN B., GÉRARD A., Chemical stimulation techniques for geothermal wells: experiments on the three-well EGS system at Soultz-sous-Forêts, *Geothermics*, vol. 38, 349–359, 2009.

PRIMAS A., *Life Cycle Inventories of new CHP systems, ecoinvent report No 20*, Tech. rep., ecoinvent center, CH-8600 Dübendorf, Switzerland, 2007.

SANER D., JURASKE R., KÜBERT M., BLUM P., HELLWEG S., BAYER P., Is it only CO2 that matters? A life cycle perspective on shallow geothermal systems, *Renewable and Sustainable Energy Reviews*, vol. 14, 1798–1813, 2010.

SANTOYO-CASTELAZO E., GUJBA H., AZAPAGIC A., Life cycle assessment of electricity generation in Mexico, *Energy*, vol. 36, 1488–1499, 2011.

SHAPIRO K., Incorporating costs in LCA, *International Journal of Life Cycle Assessment*, vol. 6, 121–123, 2001.

SHAPIRO K. Life-cycle based methods for sustainable product development, *International Journal of Life Cycle Assessment*, vol. 8, 157–159, 2003.

SINGH A., LOU H. H., YAWS C. L., HOPPER J., PIKE R., Environmental impact assessment of different design schemes of an industrial ecosystem, *Resources, Conservation & Recycling*, vol. 51, 294–313, 2007.

SPIELMANN M., BARRETO L., ERNI V., FRUTIG F., THEES O., *Life Cycle Assessment of Energy Wood Chip Supply Chains: A case study of near future supply of forest wood chips in Switzerland*, Tech. rep., Paul Scherrer Institute (PSI), Snow and Landscape Research (WSL), CH-5232 Villigen, CH-8903 Birmensdorf, Switzerland, 2007.

SPRECHER J., *Tridimensional modeling of the deep geology of the Molassic Plateau of Western Switzerland*, Master thesis effectuated at Ecole Polytechnique Fédérale de Lausanne, 2011.

STEFANIS S., LIVINGSTON A., PISTIKOPOULOS E., Minimizing the environmental impacts of process plants: a process systems methodology, *Computers & Chemical Engineering*, vol. 19, 39–44, 1995.

STEINER R., FRISCHKNECHT R., *Life Cycle Inventories of Metals Processing and Compressed Air Supply, ecoinvent report No 23*, Tech. rep., ecoinvent center, CH-8600 Dübendorf, Switzerland, 2007.

STEUBING B., *Analysis of the Availability of Bioenergy and Assessment of its Optimal Use from an Environmental Perspective*, Ph.D. thesis, Ecole Polytechnique Fédérale de Lausanne, 2011.

STEUBING B., BALLMER I., GASSNER M., GERBER L., PAMPURI L., BISCHOF S., THEES O., ZAH R., Identifying environmentally and economically optimal bioenergy plant sizes and locations: A spatial model of wood-based SNG value chains, *Renewable Energy*, vol. 61, 57–68, 2014.

STEUBING B., BALLMER I., GERBER L., MARÉCHAL F., ZAH R., An environmental optimization model for bioenergy plant sizes and locations for the case of wood-derived SNG in Switzerland, *World Renewable Energy Congress 2011* - Sweden, Linkoping, London, May 8-11, 2011a.

STEUBING B., ZAH R., LUDWIG C., Life cycle assessment of sng from wood for heating, electricity and transportation, *Biomass and Bioenergy*, vol. 35, 2950–2960, 2011b.

STEUBING B., ZAH R., LUDWIG C., Heat, electricity or transportation? the optimal use of residual and waste biomass in europe from an environmental perspective, *Environmental Science and Technology*, vol. 46, 164–171, 2012.

STEUBING B., ZAH R., WAEGER P., LUDWIG C., Bioenergy in switzerland: Assessing the domestic sustainable biomass potential,*Renewable and Sustainable Energy Reviews*, vol. 14, 2256–2265, 2010.

STUCKI S., SCHILDHAUER T., BIOLLAZ S., RÜDISÜHLI M., VOGEL F., SCHUBERT M., MAZZOTTI M., BACIOCCHI R., PAREDES G., MARÉCHAL F., GASSNER M., ZAH R., STEUBING B., THEES O., MARKARD J., WIRTH S., *2nd Generation Biogas - New Pathways to Efficient Use of Biomass*, Tech. rep., Competence Center Energy and Mobility of the ETH Domain, CH-5232 Villigen, Switzerland, 2010.

SUGIYAMA H., HIRAO M., MENDIVIL R., FISCHER U., HUNGERBÜHLER K., A hierarchical activity model of chemical process design based on life cycle assessment, *Process Safety and Environmental Protection*, vol. 84, 63–74, 2006.

SWISS FEDERAL COUNCIL, *Fiche d'information - Perspectives énergétiques 2050 - Analyse des variantes d'offre d'électricité du Conseil Fédéral*, Tech. rep., Confédération Suisse, Bern, Switzerland, 2011.

TESTER J., ANDERSON B., BATCHELOR A., BLACKWELL D., DIPIPPO R., DRAKE E., GARNISH J., LIVESAY B., MOORE M., NICHOLS K., PETTY S., TOKSOZ M., VEATCH R., *The Future of Geothermal Energy - Impact of Enhanced Geothermal Systems (EGS) on the United States in the 21st Century*, Tech. rep., Massachusetts Institute of Technology, Cambridge, Massachusetts, USA, 2006.

TOCK L., GASSNER M., MARÉCHAL F., Thermochemical production liquid fuels from biomass: Thermo-economic modeling, process design and process integration analysis, *Biomass and Bioenergy*, vol. 34, 1838 – 1854, 2010.

TURTON R., BAILIE R., WHITING W., SHAIEWITZ J., *Analysis, Synthesis and Design of Chemical Processes*, Prentic Hall, New Jersey, 1998.

ULRICH G.-D., *A Guide to Chemical Engineering Process Design and Economics*, Wiley, New York, 1996.

URBAN R., BAKSHI B., GRUBB G., BARAL A., MITSCH W., Towards sustainability of engineered processes: Designing self-reliant networks of technological-ecological systems, *Computers and Chemical Engineering*, vol. 34, 1413–1420, 2010.

VON BLOTTNITZ H., CURRAN M. A., A review of assessments conducted on bio-ethanol as a transportation fuel from a net energy, greenhouse gas, and environmental life cycle perspective, *Journal of Cleaner Production*, vol. 15, 607–619, 2007.

VON SPAKOVSKY M., FRANGOPOULOS C., A global environomic approach for energy systems analysis and optimization (part I), *Proceedings of Energy Systems and Ecology*, vol. 93, 123–132, 1993a.

VON SPAKOVSKY M., FRANGOPOULOS C., A global environomic approach for energy systems analysis and optimization (part II), *Proceedings of Energy Systems and Ecology*, vol. 93, 133–144, 1993b.

WARK K., WARNER C., DAVIS W., *Air Pollution, its Origin and Control*, Addison-Wesley, Menlo Park, California, 1998.

WERNER F., ALTHAUS H.-J., KÜNNIGER T., RICHTER K., JUNGBLUTH N., *Life Cycle Inventories of Wood as Fuel and Construction Material, ecoinvent report No 9*, Tech. rep., ecoinvent center, CH-8600 Dübendorf, Switzerland, 2003.

WORKING GROUP PDGN, *Programme cantonal de développement de la géothermie à Neuchâtel: Rapport final*, Tech. rep., Laboratoire Suisse de Géothermie - CREGE, CH-2000 Neuchâtel, Switzerland, 2010.

WYBORN D., DEGRAAF L., HAHN S., Enhanced geothermal development in the cooper basin area, South Australia, *Geothermal Resources Council Transactions*, vol. 29, 151–155, 2005.

ZAH R., BÖNI H., GAUCH M., HISCHIER R., LEHMANN M., WÄGER P., *Oekobilanz von Energieprodukten: Oekologisches Bewertung von Biotreibstoffen*, Tech. rep., Swiss Federal Laboratories for Materials Testing and Research (Empa), CH-9014 St. Gallen, Switzerland, 2007.

Nomenclature

Abbreviations

1F   Single Flash system
2F   Double Flash system
CFB   Circulating Fluidized Bed
CGC   Cold Gas Cleaning
CH   LCI dataset valid for conditions of Switzerland
CHP   Combined Heat and Power
EGS   Enhanced Geothermal Systems
DE   LCI dataset valid for conditions of Germany
FICFB   Fast Internally Circulating Fluidized Bed
FU   Functional Unit
GIS   Geographic Information Systems
GLO   LCI dataset valid for global conditions
HDR   Hot Dry Rock
HGC   Hot Gas Cleaning
HT HP   High Temperature Heat Pump
IPCC   Intergovernmental Panel on Climate Change
ISO   International Organization for Standardization
LCA   Life Cycle Assessment
LCI   Life Cycle Inventory
LCIA   Life Cycle Impact Assessment
LT LP   Low Temperature Heat Pump
MILP   Mixed Integer Linear Programming
MINLP   Mixed Integer Non-Linear Programming
MOO   Multi-Objective Optimization
MSWI   Municipal Solid Waste Incineration
NG   Natural Gas (fossil)
NGCC   Natural Gas Combined Cycle
NOx   Nitrogen Oxides
ORC   Organic Rankine Cycle (single-loop)
ORC-2   Organic Rankine Cycle with two evaporation levels
ORC-d   Organic Rankine Cycle with an intermediate draw-off
ORC-s   Organic Rankine Cycle operating at supercritical conditions
OW   Organic Waste
pFICFB Pressurized Fast Internally Circulating Fluidized Bed
PM   Particulate Matter
PSA   Pressure Swing Absorption
RER   LCI dataset valid for conditions of Europe
RME   Rapeseed Methyl Ester
SNG   Synthetic Natural Gas
UBP   Umweltbelastungspunkte (environmental load points)

UCTE   Union for the Coordination of the Transmission of Electricity
WWTP Wastewater Treatment Plant

Roman letters

A	Functional parameter related to the size of process equipment
B_b	Product or energy service to be supplied
b_b	SpeciÞc quantity of a product or service produced by a unit
c	Specific cost of a resource, service or environmental tax associated [€]
c	Correction factor related to process equipment type [-]
CI	Investment cost [€]
CO	Operating cost [€]
$\dot{E}$	Electrical power [kW$_e$]
Em	Emission or extraction of an elementary ßow of the LCI
em	Specific emission or extraction of an elementary ßow
f	Utilization factor of a unit embedded in the superstructure, [-]
FU	Functional unit quantity
h	Specific enthalpy [kJ/kg]
I	Impact over full life cycle
IC	Impact associated with construction
IE	Impact associated with end-of-life
IO	Impact associated with operation
ir	Interest rate [-]
k	Scaling exponent
l_{wf}	Yearly specific losses of working fluid in binary cycles [-]
M	Mass [kg]
$\dot{m}$	Mass flow rate [kg/s]
n_w	Number of geothermal wells [-]
P	Pressure [bar]
$\dot{Q}$	Thermal power [kW$_{th}$]
$\dot{Q}_{max}^{-}$	Design size of district heating in CHP system using an EGS [kW$_{th}$]
r	Recycling ratio associated with a process equipment type [-]
R_{an}	Annual revenue generated by system [€]
R$_k$	Cascaded heat from the temperature interval k to the lower ones [kW$_{th}$]
r_o	Yearly operation rate of the system [-]
R_r	Resource consumed
r_r	Specific consumption of a resource
s_s	Material source or sink
s	Success factor in achieving EGS construction [-]
T	Temperature [¡K]
T_a	Temperature of the cold source [¡K]
T_{lm}	Log-mean temperature [¡K]
t_p	Time associated with independent operation period p [h]
t_{pb}	Payback period (or return on investment) [yr]
t_{yr}	Lifetime of installation or of system [yr]
V	Total quantity of a LCI element
v	Specific quantity of a LCI element
x_{CH_4}	CH$_4$ recovery in CO$_2$ removal for SNG production [-]
x_d	Decision variables of the non-linear master problem
z	Targeted exploitation depth of EGS [m]

Greek letters

Δh^0	Lower Heating Value [MJ/kg]
$\Delta \tilde{h}_r^0$	Standard heat of reaction [kJ/mol]
$\frac{\Delta T}{\Delta z}$	Geothermal gradient [¡K/m]
δI	Relative reduction in the impact [-]
ϵ	Product yield [-]
η	Exergy efficiency of the geothermal conversion system [-]
ρ	Density [kg/m^3]
ω	Wood humidity after drying stage for SNG production [-]

Subscripts

an	Value calculated on an annual basis
b	Product layer
c	Value related to construction
cap	Value calculated per capita
CO_2	Value expressed in CO_2 equivalent emissions
daf	Value calculated on a dry, ash-free basis for SNG production
DH	Value related to district heating demand
dr	Value related to drying for SNG production
$e^{+/-}$	Value related to electricity input/output
e	Value related to end-of-life
EGS	Value related to EGS
ext	Value related to extracted water for EGS operation
f	Value related to flash systems for EGS operation
gas	Value related to gasification for SNG production
I	Value related to the environmental impact
i	Extraction or emission of LCI vector
in	State variable at the inlet of a unit
$init$	Initial value or quantity
inj	Value related to injected water for EGS operation
j	Element of the LCI
k	Temperature interval of heat cascade
l	Impact category
m	Material composing a type of process equipment
max	Maximal value of the given range
$memb$	Value related to membranes for SNG production
$meth$	Value related to methanation for SNG production
min	Minimal value of the given range
o	Value related to operation
out	State variable at the outlet of a unit
p	Independent operation period
pyr	Value related to pyrolysis for SNG production
r	Resource layer
ref	Reference dataset or value
s	Material source or sink layer
SNG	Identifier of a quantity related to the SNG output
tb	Value related to a turbine for EGS operation
t	Technology embedded in the units of the superstructure
th	Value related to thermal energy
tor	Value related to torrefaction for SNG production

tot	Total quantity or value over full life cycle
u	Unit embedded in the superstructure
w	Value related to the wood input
wf	Value related to working fluid of a binary cycle
z	Value related to targeted construction depth of EGS

Superscripts

$+$	Material or energy flow entering the system
$-$	Material or energy flow leaving the system
$\cdot$	Value expressed as a flow rate, as defined in Borel and Favrat (2005) [unit/s]
FU	Value brought back to the functional unit